Carl Alwin Schenck in 1951

CARL ALWIN SCHENCK

Cradle of Forestry in America

The Biltmore Forest School, 1898–1913

EDITED BY OVID BUTLER

INTRODUCTION BY STEVEN ANDERSON
PRESIDENT, FOREST HISTORY SOCIETY

Published by the Forest History Society
Durham, North Carolina

In cooperation with the
Cradle of Forestry in America Interpretive Association
Brevard, North Carolina
and the Forest Service History Program

1998

Library of Congress Cataloguing-in-Publication Data
Schenck, Carl Alwin, 1868–1955
[Biltmore story]
Cradle of forestry in America : the Biltmore Forest School, 1889–1913 /
Carl Alwin Schenck : edited by Ovid Butler : introduction by
Steven Anderson.
p.   cm.
"First published 1955 by the Minnesota Historical Society as
The Biltmore story. Reprinted 1974 by the Forest History Society and
the Appalachian Consortium, as The birth of forestry in America"
—T.p. verso.
Includes bibliographical references and index.
ISBN 0-89030-055-0 (pbk.)
1. Schenck, Carl Alwin, 1868–1955. 2. Biltmore Forest School—History.
3. Foresters—United States—Biography. 4. Forests and forestry—
United States—History. I. Butler, Ovid, b. 1880.
II. Title.
SD129.S3A3 1998   98-7308
634.9'0973—DC21   CIP

Cover illustration: Biltmore students and their first school house
Cover design: Teresa Smith Perrien

INTRODUCTION

ANY VISITOR TO THE Cradle of Forestry will sense the significance of early American forestry efforts on what was the Biltmore forest in the 1890's. George Vanderbilt hired Carl Alwin Schenck in 1895 to manage the forest on the growing estate. At the time, he was one of three trained foresters in the United States, joining Gifford Pinchot and Bernard Fernow. All three were European trained since no forestry schools yet existed in America, a point that was well recognized as the early conservation movement developed. Through this similar training all three held a common belief that in order to be successful American forestry could not follow European standards of the time; it must meet the needs of the local population. Beyond this and their obvious love and dedication to the forest, their philosophies diverged and led to serious chasms that affected the development of professional forestry and its application on both public and private lands.

In this story, Carl Alwin Schenck traces the early application of scientific principles to forestry in the United States. Having no benefits of previous research he struggled to produce positive results from nursery work, cone collecting, seed extraction, logging and fish hatcheries. While his attempt to show that "conservative" forestry could pay on the Biltmore estate was a discouraging failure, it sowed the seeds for what was to become known as sustained-yield in the 1920's, multiple use in the 1960's and sustainable forestry in the 1990's.

As Dr. Schenck took on technical help he called "apprentices", the best of these early forest rangers badgered him with a myriad of questions about the forest and the idea for a technical forestry school was born. The Biltmore Forest School was established as a school in practical forestry. Within a few weeks an undergraduate forestry program was established at Cornell University and was followed by a graduate program in 1900 at Yale. Again, philosophies differed. Pinchot and Fernow were proponents of a broad foundation of theoretical knowledge for future foresters while Dr. Schenck prepared his students for immediate employment in the woods. The same quandaries of the proper mix of theoretical and practical education for professional foresters still perplex us today. In the end, the Biltmore Forest School was ahead of its time, occurring in advance of industrial forestry and graduated its last class in 1913. Schenck's contribution to the development of the fledgling profession of forestry in the United States, however, was undeniable. Out of more than 365 students, 300 completed the required courses and more than half went into forestry work. One became Associate Forester of the United States under Pinchot; four students became District or Regional Foresters; twenty became forest supervisors or deputies; and twelve became state foresters. Others of "Schenck's boys" went into wood preservation, tree surgery, forest surveying, private timber companies, farming and international forestry consulting.

Eventually, the U.S. Forest Service purchased nearly 80,000 acres of the Biltmore estate from Mrs. Edith Vanderbilt. The Cradle of Forestry in America on the Pisgah National Forest was proclaimed a historic site by an act of Congress in 1968. Today, visitors to this forest legacy have the opportunity to celebrate a centennial of conservation. The Cradle offers guided tours along two trails, provides information about the evolution of scientific forest management through interactive exhibits, allows glimpses of Appalachian culture, and reinforces the benefits of professional forestry.

To visitors of the Cradle in the Pisgah National Forest, the Biltmore estate, or to general readers, the colorful memoirs in this book will provide wondrous tales of managing George Vanderbilt's Biltmore Forest in the backdrop of the Appalachian Mountains and society. It is must reading for the forester and forestry student as it provides insights to the development of the profession and lessons that are pertinent to forest management issues today. The historian will gain insight into the development of the early conservation movement in America. Schenck's vision, energy, and enthusiasm for training young foresters, a century ago, gives all of us a remarkable legacy.

STEVEN ANDERSON, *President*
Forest History Society

Editor's Preface

"THE BILTMORE STORY," here told in detail for the first time, is the story of the first attempt to apply the scientific principles of forestry to the woodlands of the United States. The author, Carl Alwin Schenck, a German forester, was the leader of that historical undertaking, which formed one of the most colorful chapters in the long struggle of forestry to set its roots in American soil. This account is taken from his unpublished memoirs of the years he spent in America as forester for the Biltmore Estate and as founder and director of the Biltmore Forest School, the first school of forestry in the United States.

Dr. Schenck today is the last living representative of the group of dominant leaders who pioneered forestry in America through its dawning years in the last century. His story dates back to the early 1890s, when the late George W. Vanderbilt embarked upon the building of his famous Biltmore House near Asheville, North Carolina, and the acquisition of a land estate commensurate with the architectural grandeur of the mansion — an estate that eventually exceeded 120,000 acres of mountain land. Vanderbilt announced, as one of his enterprises for the development of these extensive holdings, that the woodland areas would be managed and preserved by the application of scientific principles of forestry.

iii

At that time forestry in the United States was a theory, if not an abstraction, to the general public. Forestry was not practiced anywhere in the forests of America, and its practical application to the woodlands of the United States was a subject of apathetic debate in an atmosphere of public ignorance, doubt, and hostility. Vanderbilt's proposal therefore fell upon the public mind as a daring and startling venture, one that challenged the advocates of forestry to make good what they had been preaching and one that held national attention watchfully to the progress of forestry for some fifteen years.

When Vanderbilt began his experiment, there were only two trained foresters in the United States. Both were European trained. One was an American, Gifford Pinchot; the other was German-born, Bernhard Eduard Fernow. When Schenck arrived in America in 1895, he thus became the third forester in the United States. The stories of the other two have been told — Pinchot's in his autobiography, *Breaking New Ground,* and Fernow's in Andrew Denny Rodgers' *Bernhard Eduard Fernow.* In *The Biltmore Story,* Schenck gives his impression of the eventful years when the three men were contemporary leaders in establishing the practice of forestry in the United States. All three were men of forceful personality, strong convictions, and indefatigable energy.

Schenck's account of the period is much more than a story of forestry. It is a vivid description of life at Biltmore as he saw and lived it during the heyday of the Biltmore dynasty; of contacts with all manner of people, from the mountain hillbillies of North Carolina to the president of the United States, Theodore Roosevelt; and of frank appraisals of American conditions and people. Of special interest is his founding of the first forest school in this country and its later odyssey across America and through foreign lands.

Work began at the Biltmore Forest School in the early fall of 1898. A few weeks later the New York State College of Forestry was opened in connection with Cornell University, and in 1900 Yale University's Forest School came into being through a gift of $150,000 from the Pinchot family — Mr. and Mrs. James Pinchot and their sons Gifford and Enos. These were the first three forestry schools, and, curiously, each presented a different type of training. Yale's school was a graduate school offering a two-year course leading to a master's degree in forestry; the Cornell school, during its brief existence, had a four-year college curriculum in forestry; and Biltmore, which was open to high school graduates with lumbering experience, offered a one-year course of intensive lectures and field work in practical forestry and lumbering.

Yale Forest School was created to provide professional foresters for the government's Division of Forestry. Schenck, at Biltmore, wanted to train foresters for private industry. He hoped that lumbermen would send their sons to him for practical forestry training which they could put to use on their own timberlands. Emphasis on government forestry and on forestry as a profession requiring a broad college background and training in the basic sciences worked to the disadvantage of Biltmore, however. The lack of a permanent school home after Schenck's dismissal by Vanderbilt in 1907, a decrease in enrollment, and the failure of public acceptance of his theories all combined to discourage Schenck, and in the fall of 1913 he disbanded the school and returned to Germany.

The following spring war broke out and Schenck was recalled to the German army. He served as a lieutenant on the Eastern front and was wounded in action. After the war he returned to the practice of forestry in his native Darmstadt. During several winters in the 1920s he was a guest lecturer at the University of Montana at Missoula, and he also lectured at various other forestry schools. He took American forestry students back to Europe with him to tour the forests in the manner of Sir Dietrich Brandis and Sir William Schlich's tours in the 1880s and 1890s. During World War

II he lived quietly at Darmstadt, and after the war he co-operated with the American authorities in setting up forestry and relief programs.

In 1952, at the age of eighty-four, Dr. Schenck returned to the United States. From coast to coast he traveled, speaking at banquets and taking part in tree-farm and memorial-grove dedications throughout the country. He visited Biltmore and the homes of his Biltmore "boys" — many of whom had become grandfathers.

Schenck completed his memoirs before his last trip to the United States. He tells "The Biltmore Story" as he remembers it. The story has never been adequately recorded by the forest historian in the light of adverse conditions prevailing at the time, nor has the influence of Schenck's work and that of the students of his beloved Biltmore Forest School, who formed the vanguard of American trained foresters and helped to pioneer the profession through its formative years of reality.

Here is a field that invites thorough scholarly research and impartial evaluation. Schenck's own papers and those of the Biltmore Forest School are being collected by the American Forest History Foundation of the Minnesota Historical Society. Much information on Schenck and on Biltmore may be obtained from the Pinchot Papers in the National Archives and the Library of Congress.

This is Dr. Schenck's story, told in his own words. The part of the editor in making his account available has been merely that of extracting those parts of his memoirs that seemed most germane to the Biltmore forestry adventure and to the author's related activities, and of editing them into a story of chronological continuity. It is a story that has been too long untold as an essential chapter in the annals of early American forestry.

OVID BUTLER

Chevy Chase, Maryland

An Appreciation

PLATO IS CREDITED WITH SAYING, "Citizens, you are all brothers, yet God has framed you differently." These thoughtful words of long ago are applicable, in a very real sense, to Dr. Carl Alwin Schenck.

Whoever knows Dr. Schenck intimately will promptly affirm that he was not fashioned from a common mold. He is unquestionably one of the most individualistic, distinctive, and colorful personalities that has thus far appeared on the leadership scene of American forestry.

My personal contacts with Dr. Schenck covered a time span of forty-one years, five months, and fifteen days. We first met at Tupper Lake, New York, on April 20, 1910, that being the day that I enrolled as a student of the Biltmore Forest School. We last saw each other on October 5, 1951, the occasion of his departure from my home in Syracuse, New York, for an annual meeting of the American Forestry Association at Jefferson, New Hampshire. In a printed report on this meeting it is recorded that it "was an inspiration to the young forestry students to see the venerable but still dynamic Dr. C. A. Schenck stride about the lobby of the Waumbeck Hotel with an energy that belied his eighty-three years, and to hear the booming resonance of his voice."

As one of his students of some forty years ago, I remember

Dr. Schenck as a tall, erect, well-groomed, "handle-bar"-mustached teacher, semimilitaristic in appearance and action, unbounded in physical and intellectual vigor, and contagiously enthusiastic and impressive in lecturing to his "beloved boys."

Anyone who has ever been closely associated with Dr. Schenck will recall that he was an indefatigable worker. Day after day he would lecture several hours in the morning, conduct full afternoon field trips, and then spend his evenings, often far into the night, preparing additional lectures, reviewing and grading student diaries, appraising forest working plans, writing textbooks, corresponding with past and prospective students, and fulfilling many and various other responsibilities connected with the operation of an active forestry school. There just seemed to be no limits to his physical and mental efforts. But, irrespective of the length of his day or the drain on his energies, the next morning he showed up on time, attractively attired and in buoyant spirit, ready to tackle another full day of work.

Dr. Schenck was in no sense a specialist. Rather, he was a confirmed generalist, and was proud of it. He taught all the forestry subjects offered at Biltmore, including forest protection, silviculture, forest mensuration, forest utilization, forest management, forest policy, and forest finance. While he was not a specialist in the modern sense, he did continually and with compelling emphasis stress the importance of the fiscal and other business aspects of forestry. When doing this he was at his very best as a teacher.

Dr. Schenck was in no sense a modest or timid teacher. Instead, he was daring and dynamic. He had an abundance of self-confidence, deep and enduring devotions, strong convictions, and aspiring personal objectives. He was at his best when somewhat aroused. It was then that his arms moved, his eyes sparkled, and his thoughts glittered. On these occasions, and in quite a number of instances they were truly memorable occasions, there flowed forth from his lips word messages that were much more than an addition of his multiple talent. Instead, these special-occasion messages were a multiplication of his many talents, and it was then that he had no peer as a teacher of forestry.

Dr. Schenck was a man of broad interests and superior attainments. His teachings were not limited to forestry. Admixed with his forestry messages were lessons in art, literature, music, history, religion, and world affairs. In classroom and in field, he discussed not only the art of forestry but also the art of living. This was especially true in his later years, when he exhorted his students to live buoyantly and hopefully, but also to be sure to discover and achieve the rewards of a good and fully contributing life. And he always warned against taking things too easy, and especially against being too busy to think.

To his everlasting credit, Dr. Carl Alwin Schenck possessed that rare quality which characterizes the superior teacher who can make education remain after all that was learned has been forgotten.

JOSEPH S. ILLICK, *Dean Emeritus*
College of Forestry, State University of New York
Syracuse, New York

Table of Contents

1. BACKGROUND FOR A NEW WORLD 1

2. SIR DIETRICH BRANDIS 8

3. BILTMORE, U.S.A., 1895 16

4. PROFILE OF PISGAH FOREST 27

5. THE DAM ON BIG CREEK 36

6. THE CLOUDY DAWN OF PRIVATE FORESTRY 45

7. THE GERM OF A FOREST SCHOOL 55

8. THE PINKBEDS AND THE MOUNTAINEERS 63

9. A TRIP TO MINNESOTA 70

10. A MISTRESS ARRIVES AT BILTMORE 77

11. RANDOM JOURNEYS 86

12. SOME NOTABLES WHO CAME TO BILTMORE 97

13. FORESTERS, BEWARE! 103

14. THE YEAR OF UPTURN 111

15. THE PARTING OF WAYS WITH PINCHOT 117

16. A PORTENT OF COMING EVENTS 123

17. REVERSES AT BILTMORE 131

18. A VISIT FROM PETER THOMSON 142

19. "TRY TO SELL PISGAH FOREST" 151

20. THE BILTMORE FOREST FESTIVAL 160

21. LAST DAYS AT BILTMORE 167

22. THE ODYSSEY OF THE BILTMORE FOREST SCHOOL . . 178

23. A FAREWELL MESSAGE TO MY BILTMORE BOYS . . 201

     SOME PUBLICATIONS BY THE AUTHOR 211

     INDEX 213

List of Illustrations

CARL ALWIN SCHENCK IN 1951 . . . . *Frontispiece*

SIR DIETRICH BRANDIS ABOUT 1900 16

LINDENFELS, GERMANY 17

SCHLICH AND HIS FORESTRY STUDENTS, 1892 17

BILTMORE HOUSE IN THE EARLY 1900S 32

GEORGE W. VANDERBILT 32

FREDERICK L. OLMSTED 32

LOOKING TOWARD THE PINKBEDS OF PISGAH FOREST . . 33

BILTMORE STUDENTS AND THEIR FIRST SCHOOLHOUSE . . 48

THE SPLASH DAM ON BIG CREEK 48

SAWYERS AT WORK IN BIG CREEK VALLEY 49

LOGS IN BIG CREEK AWAITING SPLASHING 49

DR. AND MRS. SCHENCK IN PISGAH FOREST 64

THE SCHENCK SUMMER HOME IN THE PINKBEDS . . . 64

LOOKINGGLASS ROCK 65

CHESTNUT WOOD CHUTE IN PISGAH FOREST 80

LOOKINGGLASS CREEK 80

BRANDON HILL PLANTATION 81

WICKER FENCES TO CONTROL EROSION ON BILTMORE FIELDS 96

NARROW GAUGE RAILROAD IN PISGAH FOREST 96

THREE DAY CAMP, PISGAH FOREST 97

BUCKSPRING LODGE, MOUNT PISGAH 97

A HIGH-WHEELER IN THE MICHIGAN WOODS 128

OX TEAMS AT WORK IN PISGAH FOREST 128

DR. SCHENCK IN GERMAN RIDING COSTUME 129

MCGIFFERT LOADER 144

LIDGERWOOD LOGGING MACHINE 144

ROOSEVELT AND PINCHOT 145

EXPERIMENTAL STAND OF DOUGLAS FIR IN
    HEIDELBERG, GERMANY 176

BILTMORE SPECIAL TRAIN NEAR PORTLAND 177

BILTMORE STUDENTS OBSERVING WESTERN LOGGING
    METHODS 177

BILTMORE FOREST SCHOOL CLUBHOUSE 192

SUNBURST, NORTH CAROLINA 192

FOREST PLANTATIONS, BILTMORE 193

Dedication

To the young foresters of the world, in the hope that life, love, and forestry will prove to be the same challenging inspiration to them as they have been to me.

Carl Alwin Schenck

CHAPTER 1

Background
for a New World

IN MY EARLY YOUTH in Germany I chose forestry for my career. Little did I suspect then or during my years of preparation what an eventful life my choice held in store for me. Instead of following my profession in the fatherland, where forestry was well established, fate carried me to a new world — the United States of America — where at that time the practice of forestry did not exist and the word "forestry" meant to most people merely a vague and newfangled idea.

The year of my arrival in New York was 1895. In my coat pocket I carried a cablegram of employment as "forester" from one of the wealthiest and most prominent citizens of the United States, George W. Vanderbilt, whose famous estate "Biltmore," near Asheville, North Carolina, was then in the process of development. What need had he of a German forester? Vanderbilt had many idealistic plans for his castle in the mountains. One of them was forestry, and as his forester I was expected to set a pattern or example of profitable sustained-yield forest management on thousands of acres of mountain land which were to form a part of his grand estate.

More than fifty years have elapsed since I stepped from the boat in New York. During the first decade and a half of those years, I saw and was an active participant in the slow borning

1

of forestry in the United States. Now in my eighty-third year, I am here setting down recollections, refreshed by diaries faithfully kept, of experiences, problems, and people that crowd my memory. They are offered merely as a contribution to contemporary history of the colorful period out of which American forestry has emerged to its present magnificent stature.

Why did Vanderbilt engage me, a young German forester, for the pioneer task of establishing forestry on privately owned lands in the United States? I had never been in the country, and my knowledge of its forests, its people, its language, its customs, and its economy were meager almost to the vanishing point. What preparations and qualifications, then, did I have for this quixotic call to a new world?

The answers to these questions, personal though they may seem, form an essential background to my encounters with realities in a strange country. My adolescent intentions to become a forester were encouraged by my parents, mainly, I suspect, because I was the sickliest of five boys and they saw in forestry a remedy for my infirmities. After graduating from the Institute of Technology in Darmstadt, Germany, at eighteen years of age, I was not allowed to go to a university for a postgraduate course in forestry but was compelled to spend another year between Darmstadt, where I was enrolled as a student of botany, and Lindenfels, a fresh-air resort where my parents had a summer place.

Now it happened that in Darmstadt I fell in love with an English girl who was getting an education there, and I began to study English with her as a teacher. For her sake I learned to recite by heart almost all of Shakespeare's *King Richard II*. My love affair ended, however, in the spring of 1887, when I was sent to the forestry school at the University of Tübingen. The dean, Tuisko Lorey, was a Darmstadtian, a friend of my family and then famous as the author of the first encyclopedia of forestry ever written.

Let me admit that I did not attend his lectures. Members of the Suevian students' club in Tübingen, of which I was one, spent their time mostly in dueling, drinking, dancing, and other sports. In one of these escapades I was almost drowned in a small lake in the Black Forest, and as a result I was sent home with a severe affliction of my lungs and was forced to spend the rest of the year at a tuberculosis institution in the Taunus Mountains. There I had a good time with an aristocratic T. B. club called "The Indians," whose members were equipped with wigwams, bows, and arrows. I was baptized by the Indian chief as "Le Lévrier Ingénieux," and by my stunts and "Indian songs" I believed I was adding a good deal to the general hilarity of the invalids.

My next move was to the forest school of the University of Giessen, one of the earliest schools of forestry to be combined with a university. Lordy! two years had elapsed since my graduation in Darmstadt, and I had lost, by sheer idleness, the two best formative years of my life. I was disgusted with myself, and from that time on there was no student at the university more industrious than I. At Giessen a capital teacher, Dr. Richard Hess, was dean. A modernist, Dr. Karl Wimmenauer, was instructor in such refinements as forest finance, and our teacher in physics was — imagine! — Wilhelm Konrad Roentgen, who a few years later discovered his famous Roentgen rays. Always in a blue suit, Roentgen was the best-dressed man, and I might add the only well-dressed man, among the professors. My teacher in agriculture was a grandson and the namesake of Albrecht Thaer, founder of the first school of agriculture in Germany. He spoke English well, after a long residence in Great Britain. Actually, my knowledge of *King Richard II* impressed him so much that I later passed my farming examinations with distinction.

My lungs continued in such poor condition, however, that I was declared unfit for army service, and in Germany physical fitness was a *conditio sine qua non* for an appointment in the forestry service. What sense was there in all my preparation for forestry if I could not obtain a position after passing the prerequisite examinations? As a possible alternative career, I studied law along with my studies of forestry. A law student could be admitted for the state examinations in Germany at that time if he could prove that he had attended a university for three years. There was no other requirement. Fortunately, during my two years of loafing just ended, I had been officially enrolled as a student at Darmstadt Tech and at the University of Tübingen. Adding the two official years thus spent to one more year of real studies in forestry, farming, physics, chemistry, geology, and law at the University of Giessen, I could qualify for, and I eventually passed, the law examinations! True, I barely passed, but that was all I needed as a safety outlet should I fail to be acceptable for forestry. Let me interject here that in my later life in America my small knowledge of the rudiments of law was of the very greatest importance to me. Three cheers for Emperor Justinian and his universal corpus juris, which formed the basis of law in America, in England, and in Germany! Meanwhile, I also passed with honors the first examinations in forestry where a knowledge of chemistry, physics, mathematics, geology, botany, and zoology had to be shown by the candidate.

While I was at the University of Giessen an event that influenced my whole forestry career occurred. In June, 1889, the forest school of the university was honored with a visit by one Sir Dietrich Brandis, accompanied by some forty graduates of the Royal Indian Engineering College at Cooper's Hill, near Windsor, England. Who was Sir Dietrich Brandis? It was Brandis, we were told, who had first introduced forestry to the English commonwealth of nations. As inspector general of forestry in India, he had spread the practice of forestry all over its many provinces, aided since 1866 by William Schlick. Schlick, a native, like myself, of the German

state of Hessen, succeeded Brandis in 1883 when the latter
left India to live in his parental home city of Bonn, Germany,
for the remaining twenty-four years of his life.

Who were the forty graduates of Cooper's Hill? They had
been drilled at that institute by Schlich, who had recently
retired from his Indian position of inspector general to be a
staff teacher for the new Imperial Indian Forest Service. The
forty candidates for that service were fine young men who, it
seemed to me, had selected forestry for a career, not because
of enthusiasm for the profession, but because they had failed
to pass their examinations for the Indian civil administration
or those for the English army and navy. Cooper's Hill and
England, at that time, were — and are to the present day — a
place and a country where there existed neither laboratories
for forestry nor object lessons. The students obtained, as they
do nowadays at Oxford and Cambridge, merely a book
knowledge of forestry. To overcome this handicap, the candidates for the Indian Forest Service, on the suggestion of
Brandis and Schlich, were required to spend a year with
various German supervisors of forestry touring the woods of
Germany, Switzerland, and France. On these tours they were
instructed as well as examined by Sir Dietrich.

And now, in June, 1889, this English-Indian delegation
came to Giessen to inspect the experimental forests and the
city forests near by. To fraternize the occasion, the German
students staged a huge *Bier-Kommers* for the English boys.
It was a joyous drinking spree, with informal speeches, songs,
and fun. Naturally, an orator was needed by the Germans to
welcome their English guests. Since my rudimentary knowledge of English was known among my fellow students, I was
selected for the post. Sir Dietrich was present. He must have
liked my address of welcome, because when the affair was
over he took me aside and asked me to accompany him, as his
assistant, on the tour through the woods of western Europe.
I accepted. The long summer vacation of the university was

about due. I was at liberty and overjoyed at this unique chance to see the woods about which my professors, without having seen them, were lecturing.

On our visits to the various European woods, Sir Dietrich was in the habit of taking "sample plots" with the help of the students. He himself used a surveyor's staff, six feet long and striped white and red, for a walking cane. The students carried calipers, measuring tapes, and staffs, by which the corners of rectangular sample plots were designated. While Sir Dietrich, with a crew of four men and with his "Brandis clinometer," measured the heights of some twenty trees within the plot, I myself supervised three caliper men who were traversing parallel avenues within the plots, all some twelve feet wide, and framed by ordinary white tapes dragged meanderingly along the ground by three of the students. This "Brandis system" of cruising was capital and it was quick. There was one trouble about it which Sir Dietrich overlooked. All measurements were taken in the metric system, which had no meaning for the minds of his English followers. For me, German born, the game was fascinating indeed. It dawned on me that Sir Dietrich, while he looked and dressed as an Englishman and conducted the regular Sunday services for the assembled group after the rites of the English High Church, had remained German to the core.

Sir Dietrich kept me busy day and night. Frequently I worked with him late into the night until the clock struck eleven, and thereupon he asked me to have ready the next morning an elaboration, which it took me three or more hours to complete. In addition, to augment his own historical studies of the Spessart Forest or of the Black Forest, I submitted to him essays on their histories based on my own studies made at various libraries. While he was cordial with local forest authorities, he disliked to sit up late with them over a bottle, or shall I say six bottles, of wine. Many were the evenings, however, when he could not escape from German convivials. Fortunately I had nothing to do with the financial accounts of the Brandis tours, accounts which evidently took much of the punctilious leader's time.

There is another matter that is deeply impressed on my memory. The few Hindus, Parsees, and Brahmans among the students were treated worse than dogs by their English fellow students. No Englishman was an intruder in India, but every native who went to Cooper's Hill was unwelcome. Among the English was the son of an English administrative officer of high standing who had married an Indian princess. This son was treated by the full-blooded Englishmen with disdain. No wonder that the Indians did not mince their words when talking to me about the English! Sir Diętrich seemed not to have eyes nor ears for these discrepancies and animosities. And I myself? Let me confess that I sided with the English!

CHAPTER 2

Sir Dietrich Brandis

IN EUROPE THE NAME of Sir Dietrich Brandis had become traditional among foresters everywhere. So it became in the United States as an increasing number of young men embraced and studied forestry. Indeed, Sir Dietrich might be called the "grandfather of American forestry," although he never visited the United States. Undeniably, Gifford Pinchot, Overton Price, Henry Solon Graves, Frederick E. Olmsted, Austin Cary, and Edwin M. Griffith, all of whom were important in the early forestry movement, were decisively influenced by him, as was I in particular. No man has ever left with me personally a deeper impression than has Sir Dietrich. He was an indefatigable worker, filled with the highest ideals, absolutely unselfish, a stalwart Christian. President Theodore Roosevelt, sending his portrait to him in 1904, praised Sir Dietrich's influence on forestry in the United States.

Dietrich Brandis was born in Bonn on the Rhine in 1824, the son of Christian August Brandis, a Bonn University professor of philosophy who had enjoyed the friendship of King Otto I of Greece. After completing his botanical studies in Bonn, with a degree of doctor of philosophy in his pocket, young Dietrich Brandis married an English lady, a member of the aristocracy and a connection of James Ramsey, Earl of Dalhousie, who was made governor general of India at the age of thirty-five. When Lower Burma (Pegu) was added in 1852 to the English colonial empire, Lord Dalhousie offered Brandis the position of forest superintendent in Pegu. Dalhousie's uppermost interests in Pegu were not the teak trees in its forests waiting for exploitation, but the humble natives of Pegu, a tribe known as Karens, whose monosyllabic and guttural language the new superintendent, Brandis, was quick to learn.

What sort of forestry was Brandis to introduce? His forestry was meant to keep the natives at work through logging and to provide them with needed rice fields; to satisfy the teak hunger of the merchants in Rangoon; and, notably, to maintain undiminished the volume of teak stumpage in the primeval woods. Brandis, so he told me, had traversed the woods of Pegu riding an elephant on such trails as there were, with four sticks in his left hand and a pocketknife in his right. Whenever he saw in the bamboo thickets a teak tree within two hundred feet of his trail, he cut a notch in stick number 1, 2, 3, or 4, denoting the diameter of the tree. It was impossible for European hands, dripping with moisture, to carry a notebook. At the end of the day, after traveling some twenty miles, Brandis had collected forest stand data for a sample plot four hundred feet wide and twenty miles long, containing some nineteen hundred acres. He continued his cruise for a number of months, sick with malaria in a hellish climate. Moreover, he underwent a trepanning operation, and for the rest of his life he carried a small hole filled with a white cotton wad in the front of his skull. But he emerged from the cruise with the knowledge needed for his great enterprise.

On his cruises Brandis had also studied the rate of growth of the teak trees, and he knew that, if the teak trees three or four feet through were removed, forty years would be required for the young trees one and two feet in diameter to replace them. For the reproduction of teak, a system was introduced known in Germany as "Waldfeldbau" and in Burma as "toungya." Under this system, after clear-cutting, the soil is used temporarily for farm crops — in the case of Pegu for the production of rice. The native Karens tended and fenced their rice fields, and when teak seedlings were planted with the rice they were thus protected from the ravages of wild animals, especially those of the wild elephants. This toungya, first introduced by Brandis, is continued to the present day in several provinces of the former British Indian Empire and in Indonesia.

Brandis had not studied forestry at a school of forestry, although on his travels in Europe he had seen it practiced; but no "scientific" forester could have done better than this layman. Verily, the best college of forestry should be a college on wheels, or, as in the case of Brandis, should I say, on the head of an elephant. When he retired in 1883 from active service as inspector general of forestry in India, the Indian Forest Service was fully established. A division of working plans and an Indian Forestry School had been developed under him by William Schlich, whom Brandis had taken to India in 1866.

While I came to love and admire Sir Dietrich during our trips in Germany, his English followers did not wholly share my feelings, mostly because he compelled them to work and to work hard, day after day. And he forced them to write every evening in their diaries, describing the events of the day. On Saturdays the diaries were turned in to me for corrections and for such criticisms as I thought fit. The students' work under Brandis, and the grades they received, determined the province in India to which they were to be sent. The best man was sent to the northwest provinces, with the best climate; the worst was left for Madras, famous for its terrible climate. Let me confess here, however, that in my later experience as a teacher of forestry, I found that the young man who excells in scientific examinations is by no means predestined for the most efficient career in forestry. As an assistant of Brandis on his European tours, I came in touch not only with the most interesting woodlands of western Europe, but also with the leading men in forestry, among them Johann Friedrich Judeich in Tharandt, Franz Fankhauser in Bern, and Ulrich Meister in Zurich.

Sir Dietrich was interested in forestry the world over, and his interest in American forestry was intense. Had it not been for my accidental contact with him in 1889 I would not have had a chance to go to the United States and would have lived and died, at best, a professor of forestry at one of the German schools. For me his interest in forestry in the United States was of the greatest help and encouragement after I arrived in that country in 1895. Between that year and 1907, the year in which he died, I received from his pen some thirty letters, of which the following is characteristic:

BONN, DECEMBER 2, 1898

My dear Dr. Schenck:

I am much obliged for your letter of 19th ultimo; Lady Brandis also will be glad to hear from you. She is progressing fairly well, but will not return home for some time yet.

Should I ever have the pleasure of again meeting you with leisure for a good discussion, I want to point out to you the limitations of the principle, first, I think, expressed by Dr. Schlich that *good forest management is the one which complies best with the wish of the owner.*

You have my little pamphlet on Indian Forestry. In 1861, when the Government of India ordered us to open the Pegu forests to private enterprise, the Government of India undoubtedly was the legal representative of the owner of these forests. Yet the measures which the commissioner, Colonel [Arthur] Phayre and myself took at the time, had the effect of organizing matters in a manner not at that time intended by the Government of India.

Ribbentrop [1] writes to me that the present Chief Commissioner of Assam, who is the representative of the owner of the state forests in Assam, intends to open them all for the extension of tea plantations and that he hopes to thwart his intentions.

Even a private proprietor is not absolute or unfettered. There are certain considerations which are more mighty than the fleeting fancies of a rich and thoughtless man, or the greed of a man anxious to become rich.

The question is not as plain as it would at first sight appear, but it is too difficult to deal with in a letter.

Very glad you will assist Pinchot in his private forests working plan scheme. The drawback of that scheme is that there are no trained foresters to carry out the provisions of his working plans.

What you should aim at, are *preliminary* working plans, providing only for a *limited* number of years. When I left India I had hoped to carry out this idea with regard to a number of ranges.

After I left the mistake was to make elaborate plans for *long periods* and many of these plans were found to be utterly impracticable.

I hope that you and, through you, Mr. Pinchot may fall in with my idea of *preliminary working plans*. Do not attempt more than you have data for.

Your Pisgah [Forest] working plan, if there is a map with it, however rough, will interest me much. But you must not expect me to give any advice regarding the management of a forest which I do not know myself.

You speak of *conservative lumbering*. I am not sure whether I quite go with you in this. My principle is and has always been: While we are in ignorance concerning most factors affecting growth and regeneration of the trees, cut as little as possible! This, however, is not conservative lumbering.

As soon as you have prepared for Pinchot a plan according to your ideas get it printed sharp and send it [to] me with the map, however sketchy. But do not attempt to lay down any operations beyond the time *of 10 years*. That, practically,

[1] Baron Bernhardt von Ribbentrop had been inspector general of forestry in India since 1889.

is the limit in Baden and Württemberg. And it is the secret of their admirable management. Mind, my dear Schenck, the foresters of these two states attain the same end by two entirely different methods: But in the *10 year duration* of their working plans, they are as one.

For you I should be glad to have an even shorter time adopted. Five years is quite sufficient to begin with. Should you fall in with my ideas in this respect you may be the founder of a system of working plans adapted to circumstances in the U.S.

I sympathize to the fullest extent with your wish to show that under given conditions American Forestry can be made remunerative. For that, however, you ought to have the management of a forest that *can be made remunerative,* where there is still a stock of merchantable timber on the ground and where growth is fast and natural regeneration is easy.

The only places I know of where these conditions still exist are the Redwood Forests on the coast range of California and the Douglas Fir Forests in Oregon and Washington.

There may, possibly, be woods left worth managing in the Pitch Pine Forests of the South and in the white pines of the Lake country.

There are two powerful interests at work in the U.S. in favor of conservative forest management:

I. The wish of lumbermen to adopt conservative lumbering.

II. The millions of capital invested by paper pulp manufacturers in machinery.

If I do not mistake, these two interests will do more for the cause of forestry than either the federal government or the governments of the individual states. I will except Pennsylvania where, under Professor Rothrock,[2] matters seem to be progressing on correct lines.

I hope you will soon have the management of a forest worth managing. Neither Pisgah nor Biltmore are worthy of you.

With a forest that pays under good management you will

[2] Joseph T. Rothrock was lecturer on forestry at the University of Pennsylvania and secretary of the Pennsylvania Forestry Association.

be able to train the U.S. foresters of the future. In Pisgah and at Biltmore you can doubtless teach them a great deal, but the proof of the pudding is in its eating and steadily growing net revenue would make your pupils foresters at once to the backbone, because it could teach them that good forest management pays and is a reality.

When you see Pinchot or you write to him, will you say, that if I am not much mistaken I have replied to all his letters and that I now expect to hear progress in regard to his scheme.

I am now engaged with [Overton] Price in working up my notes on the forests in Baden, Württemberg, Saxony, Bavaria for publication in the U.S. I had hoped to publish them in England but there it is quite hopeless to get a publisher. Price is confident to get them published in the U.S. There will be three maps showing topography and geology of Schwarzwald, Spessart and Steigerwald and the Erzgebirge. The work has advanced considerably and I hope to see it finished, if nothing unexpected occurs. It will be a volume of 400 to 500 pages with a few other illustrations besides. Your and Dr. Schlich's notes on my papers (Saxony, Württemberg) I have been using as far as I agree.

I had always hoped to have done this work with you, but of this there is no chance I see, with the multitude of your engagements.

The book is to be a sort of a Baedeker for Americans and Englishmen wishing to study forestry in Germany.

May I ask you to assist Price in reading and correcting the proofs? That business I must entirely leave to him. Both our names will appear on the title page.

I am much inclined when the book is ready to ask Professor Sargent whether I may dedicate it to him. He is not a forester, and, while chairman of the commission may have proposed impracticable things. But by his *Silva* and by his previous work in connection with the Census and, lastly, by his editorship of *Garden and Forest* he has done much to promote good forest management in the U.S.[3]

[3] Charles S. Sargent's *Report on the Forests of North America (Exclusive of Mexico)* was published as volume 9, *United States Census*, 1880 (Washington, 1884). His *Silva of North America: A Description of the Trees which Grow Naturally in North America, Exclusive of Mexico*, was published in fourteen volumes (Boston and New York, 1891–1902).

The first step in a country with so varied a forest vegetation is to describe the species which compose the forests. And Sargent's *Silva* has placed that business upon a permanently safe footing.

Will you kindly consider this idea of mine and if you should by any chance meet Professor Sargent will you find out whether he would like it?

There is one difficulty and that is he and Pinchot have fallen out. I do not wish to vex Pinchot by such a step, and therefore write to you asking you to consider my idea from all sides and tell me your conclusions.

Whether I dedicate the volume to him or not, in the preface I shall give full vent to my feelings of respectful admiration of what Sargent has done to promote rational forestry in North America.

I hope Mrs. Schenck is well and that you are flourishing.

Believe me most truly yours,

D. BRANDIS

CHAPTER 3

Biltmore, U.S.A., 1895

WHEN SIR DIETRICH BRANDIS retired in 1891 his place as leader of the English-Indian visitors to Germany was taken by Sir William Schlich of Oxford, founder of the forestry school of the Royal Indian Engineering College at Cooper's Hill. I continued as Schlich's secretary during the years 1892, 1893, and 1894. These tours kept me busy two or three months in summer. During the rest of the time I completed my university studies, passed my school and state examinations, and obtained, early in 1895, a Ph.D. degree at the University of Giessen. The degree proved to be a good asset for me later in the United States. Foresters having a doctor's degree were at the time rare in Europe and a species virtually unknown in the United States. I had worked hard during the five years of my preparation, which were interrupted only by compulsory military service, to which every German was subject. I absolved it in the Grand Ducal Horse Artillery at Darmstadt, in which I attained the rank of a lieutenant.

In January, 1895, my uncle, Max Müller-Alewyn of Russian nationality, who had retired from the Russian consular service, invited me to spend a few weeks in Menton on the French Riviera with him and his wife, sister of my mother. During my stay with them I received a cable from the United

Sir Dietrich Brandis about 1900

Lindenfels, Germany

Schlich and His Forestry Students, 1892
Schlich, seated center, holds a light hat. Schenck is seated at the extreme right.

States reading: "Are you willing to come to America and to take charge of my forestry interests in western North Carolina?" The message was signed "George W. Vanderbilt."

I was nonplused, not knowing what to do. A brother of mine had died in the West Indies a few years before, and at his death I had promised my mother that I would not seek employment in foreign countries. Also, I was engaged to be married to a beautiful girl in Darmstadt and I was uncertain of her willingness to go to the United States. One thing was clear to me at once. The originator of this offer could be no one but Sir Dietrich Brandis; and if Sir Dietrich was willing to recommend me for a job I was sure it would be a good one.

My Russian uncle was radiant with joy. Foreign countries had for him more attraction than Germany. He had been advising me for some years to go to India or to the English colonies. And, of course, the Vanderbilt name was familiar to him, inasmuch as his investments, large as they were, consisted in part of American stocks. What was my answer to the cable to be?

Fortunately a boyhood friend of my father, an American colonel, John O. Siebert, was wintering at the time with his wife and daughter in Nice, an hour's ride from Menton. I sought his advice. Colonel Siebert congratulated me with fervor at the chance to establish forestry in the United States on the lands of a multimillionaire well known throughout the country. In the colonel's opinion, Vanderbilt's invitation opened to me a brilliant future. Thus, with his encouragement, I cabled the following answer to Vanderbilt's inquiry: "I accept your proposition subject to Sir Dietrich Brandis' approval."

In my reply I kept open the possibility of a retreat, putting the responsibility for my acceptance on Sir Dietrich, because I could not accept unconditionally against the wishes of my beloved mother and my fiancée. Fortunately my fiancée was

willing to take American risks, and my mother was persuaded to acquiesce when a leading banker and friend of the
family, Carl Parcus, gave a glowing description of western
North Carolina, which he had recently visited.

Brandis, when I visited him at Bonn, confessed that he
had proposed my name when a young American, Gifford
Pinchot, had asked him for advice. Brandis had known Pinchot personally while the latter was studying forestry at
Nancy in 1891. He told me that Pinchot was the son of
wealthy parents and a great sportsman and huntsman; that
he had worked for a short time in several German forest districts; and that he had made some early attempts in forestry
on Vanderbilt's place in North Carolina and had issued a
pamphlet descriptive of this early work to accompany a
timber exhibition which he had prepared for the Chicago
World's Fair in 1893. Sir Dietrich insisted that I could rely
on Pinchot's sincerity and help during my stay in the United
States.

Before I left Germany I had a letter from Pinchot himself.
He told me that, under American laws, a contract of employment made across the ocean was not feasible, but that I
would receive an annual salary of twenty-five hundred dollars after my arrival at Biltmore, a place already famous in
the United States. I should interject here that Pinchot had
been engaged by Vanderbilt in 1891 as consulting forester of
the Biltmore Estate to prepare and carry through a plan for
the management of the estate forests, and that he was serving
in that capacity when I arrived in the United States.

On April 5, 1895, I passed the Statue of Liberty and was
received in Hoboken at the docks of the North German
Lloyd Lines by my friends and cousins, George Merck and
his wife, residents of New York since 1893. They were overjoyed at my coming. Their son, George W. Merck, who as I
write is a director of the American Forestry Association, was
just one year old. I was taken in a horse-drawn cab, with the

help of a ferry, to New York and to their residence on Eighty-sixth Street. In the afternoon George Merck drove me in a buggy through Central Park. There were thousands of buggies, and they seemed to me more admirable than any of the trees in the park and more attractive than the entire bank of the Hudson River, including the Grant Memorial. I recall that everybody craned his neck whenever a woman riding on a bicycle was seen — apparently a new sensation for New York in 1895.

Merck was United States representative of the German chemical firm of E. Merck of Darmstadt. I cannot say that I was much impressed by the firm's New York offices and storerooms on William Street. How could I foresee at the time that from this small beginning would arise the second largest pharmaceutical enterprise in the world? Or that George Merck's infant son would become its president?

On the third day after my arrival in the United States, Pinchot came to the Merck home and picked me up for a day to be spent in New York City. He was good looking. His shining black eyes and black moustache betrayed his French origin. Brandis had told me that Pinchot's grandfather had kept a military canteen in Napoleon's army and had emigrated to the United States after Waterloo. When the grandfather had made a great success of a French restaurant on Broadway, Pinchot's father, James W. Pinchot, had married Mary Eno, the heiress of one of the largest hotels in New York City. Young Pinchot was led to forestry, it seems, by an intense love of nature acquired at the parental estate at Milford, situated in the northeast corner of Pennsylvania.

Gifford Pinchot appealed to me as the most lovable companion I could desire. To begin with, we inspected the city of New York, riding in horsecars, in cable cars, and on the new elevated railroad. In those days electric lines did not exist. We visited the American Museum of Natural History, where we saw the Morris K. Jesup collection of native American woods. O Lo! The biggest trees I had seen in the Spessart and in the Black Forest were mere babies when compared to the gigantic dimensions of American trees. In the course of our trip we visited a large store, Rogers Peet Company, where there could be had all and everything that a gentleman might require for society, for business, or for sport. I was amazed when Pinchot selected, without asking the prices, several sport suits, the finest touring shoes, some tents, and some fishing tackle. Obviously, he was well known at the store. He himself was clad in black, from neck to foot. Apparently he was in mourning; but his cheery eyes were in strict contrast with his mourning attire.

At noon he took me to his home, a patrician house in Grammercy Park. Upon entering, Pinchot introduced me to his father as George Vanderbilt's new forester and added that he had invited me to have lunch with the family. Much to my astonishment, the elder Pinchot replied, without looking at me and without giving me a hand: "No, it does not suit us today; you have to take him elsewhere." Undismayed by his father's brusqueness, Pinchot left the house with me at once and took me to his Yale Club for lunch.

Subsequently Pinchot and I met at various places and for various purposes. Queerly, the task awaiting me at Biltmore was scarcely touched in our conversations, which were restricted rather to discussions of hunting and fishing. Pinchot gave me the name of the general manager of the Biltmore Estate, Charles McNamee, and advised me to call on him at once upon my arrival. A brother of McNamee's had married a distant relative of the Vanderbilts. Seeing me off at the depot in the evening, Pinchot paid me this compliment: "Dr. Schenck, I believe you are just the right man for the position." Then he added, "You will be forester and I shall be chief forester during your term of employment." He had promised to go to Biltmore in the near future and to discuss with me all the problems on the spot. I was happy, indeed, to be able to work under him on the Biltmore Estate and under his responsibility.

On the way south I stopped in Washington, D.C., in the hope of obtaining, through the German ambassador, Theodore von Holleben, an introduction to the chief of the Division of Forestry in the Department of Agriculture. Although I was in possession of a special letter of introduction signed by the prime minister of Hessen and addressed to the ambassador, that dignitary had no time for me. Thus, after wasting a day in Washington, I made bold to knock at the door of Bernhard E. Fernow, who headed the Division of Forestry. Fernow welcomed me joyously. A German forester had come to America to take charge of first attempts at practical forestry in the United States! Here was the beginning of the fulfillment of his fervent hopes!

Fernow told me that he had studied forestry in the early seventies in the Prussian forestry academy at Muenden and that he had come to the United States in 1876 to attend the Philadelphia Centennial Exposition, after which he had become an American citizen and had devoted himself to the advancement of forestry. In 1883 he had been appointed secretary of the American Forestry Congress, an association formed in Cincinnati in that year under the influence of Baron Richard von Steuben, a Prussian forester, during a visit to the United States. The baron was a grandson of Major General Friedrich Wilhelm von Steuben, who had been George Washington's helpmate and his army organizer. President Cleveland in 1886 had appointed Fernow chief of the Division of Forestry, a small agency of uncertain status in the Department of Agriculture.

Fernow told me that his work of forest investigation, preparation of reports, and general forestry promulgation was badly handicapped by an annual budget of only twenty thousand dollars. After half an hour of conversation conducted in German, he introduced me to the secretary of agriculture, Julius Sterling Morton, who received me with open arms. Morton told me with great pride that he had been interested particularly in prairie tree plantations and that he was the originator of the Arbor Day movement in his state of Nebraska. He promised to visit me in Biltmore within a fortnight, and he seemed to be deeply interested in the Biltmore venture. After this interview Fernow took me home with him to supper. His wife was charming and hospitable. After supper I had a chance to admire Fernow's talents at the piano. When the day was over, I felt that this had been my best and most eventful day in the United States.

In Fernow's office I had met Filibert Roth, who also had come from Germany. Roth was at that time special agent and expert in timber physics in the Division of Forestry. Fernow mentioned that he had offered Pinchot also a position in his office upon the latter's return from Nancy, but that Pinchot had declined. Unless my memory fails me badly, neither Pinchot nor Fernow mentioned to me two names which, it soon thereafter dawned on me, were those of leaders in American forestry. One of them was Joseph T. Rothrock, originator of the forestry movement in Pennsylvania, secretary since 1892 of the Pennsylvania Forestry Association, and editor of *Forest Leaves*. The other was Charles Sprague Sargent, author of the monumental *Report on the Forests of North America*, director of the Arnold Arboretum in Jamaica Plain, Massachusetts, and, at the time, beginning to publish his phenomenal *Silva of North America*. He was admired, as I have related previously, by Sir Dietrich Brandis.

Sargent also had been publishing, since 1887, *Garden and Forest*, a weekly devoted to horticulture, landscape art, and forestry. Lest I forget to mention it at a better place, there were published in four issues of this weekly, beginning with June 16, 1897, four papers of mine — my very first in the

United States — bearing the title "Private Forestry and State
Forestry." My leading thought at that time seems to have
been that "to act unwisely is wiser than not to act at all" —
an observation exculpating the early mistakes by Pinchot,
Fernow, and, notably, myself.

I left Washington by the night train of the Southern Railway and arrived the next day in Biltmore. Biltmore was
nothing. Only a few buildings stood where there is now a
beautiful village. There was, however, a bus of the Kenilworth Inn in which I took refuge. The inn was a huge
affair which, it seemed to me, had more Negro waiters than
guests. At the window of my room was a coiled rope which
the waiter explained was for use if there should be a fire in
the hotel. This explanation was most reassuring to a German
guest unaccustomed to hotels built entirely of lumber.

The next morning I paid a visit to Charles McNamee,
manager of the Biltmore Estate. He received me most kindly
and told me that my office would occupy the top floor of his
own office building and that he would secure for me a good
stenographer. It was hoped that George Vanderbilt himself
could come to Biltmore and personally make the arrangements involving my future position.

The Biltmore Estate at the time was in the making. There
was a huge castle, not yet completed, to which led a railroad
two miles long. The architect was Richard Morris Hunt,
famous for his extension of the Capitol in Washington and
for the construction of the Lenox Library in New York and
the pedestal of the Statue of Liberty in New York Harbor.
Frederick Law Olmsted, equally famous for his park crea-
tions in New York and Boston, was the landscape architect
responsible for the plans, embellishment, and sites of all
buildings on the entire estate.

All these plans were based on the best map I had ever seen
— a topographic map with contour intervals of five feet, which
had been made at an expense of some thirty thousand dol-

lars! At the entrance to the estate was a brick factory and, close to the Asheville railroad depot, a furniture factory and a band-saw mill, all working for the construction of Biltmore House and the estate. Never had I seen a private enterprise of this magnitude. Imagine! Every stone used for house-building was limestone brought from Indiana over a distance of six hundred miles! Limestone, the proper material of which to build a Renaissance castle, was not available in North Carolina. I was told that two years earlier the site of the castle had been a valley, and not the promontory now filling the valley.

On the north bank of the Swannanoa River were large tree nurseries in charge of a landscape department under Chauncey D. Beadle. A dairy farm, a pig farm, and a poultry farm were in the making. Unfortunately Baron Eugene d'Allinge, who had been chief of all farm work, had just died. The Baron was an erstwhile German officer and a chance acquaintance of George Vanderbilt, and his assistant was George F. Weston, a young engineer. Apparently a knowledge of farming was not required to develop a huge farm in the United States.

Following an invitation from Mr. and Mrs. Frederick Law Olmsted, I became their guest in their summer residence on the Biltmore Estate. My sojourn with this charming family, which included a daughter, Marion, and a rattling good lad, Frederick Law, Jr., was to be of the greatest importance to me, for it was by them that I was introduced to America and to the life on the Biltmore Estate. It began to dawn on me that the idea of forestry at Biltmore had been sponsored by the elder Olmsted, my kind host who had seen forestry at its best in Germany and who saw in it an adjunct to his own profession of landscape architecture. Through his efforts the dilapidated woodlands and the farms surrounding Biltmore House were being transformed into a paradise, dissected by the best of macadamized roads. Also, Olmsted had submitted a plan, which Vanderbilt had accepted, for a grand arboretum to be created on the Biltmore Estate. The Arboretum Road, some ten miles long, which was to traverse this arboretum longitudinally, already had been constructed. In the Biltmore Nurseries, I was informed, more tree species were on hand to be planted in the arboretum than the number of species in the famous Kew Gardens near London.

I suppose Olmsted did not discover, at the time, my total ignorance of dendrology. His loving-kindness and his gentility were shining not merely in the cottage in which he lived but also on the entire Biltmore Estate. Mrs. Olmsted, the lady of the house, bright as the sun, was guiding the family and incidentally instructing me in the affairs of the Biltmore Estate and in those of its personnel. Much more critical than her husband, she was superior to him in the unreservedness of her judgment.

While a guest of the Olmsteds, I was shown through the woods by the ranger, Charles L. Whitney, whom Pinchot had brought from the Adirondack region of New York State. Whitney was a professional lumberman, some thirty-five years old, and was in charge of all logging operations at Biltmore initiated by Pinchot. These logging operations had consisted, in the main, of the removal of the large number of dead and dying trees, notably chestnuts, found all over the estate. They had been cut into fuel wood, of which thousands of cords were scattered throughout the grounds. The brickkilns on the estate, which had required large amounts of fuel wood, had been abandoned. Thus when I came to Biltmore there were no outlets nor markets for this wood.

And the woodlands themselves? The Biltmore Estate surrounding Biltmore House, approximately 7,280 acres in extent, was made up of some fifty decrepit farms and some ten country places heretofore owned by impoverished Southern landed aristocracy. Most of the cabins and houses of the former owners had been removed; and they continued to be removed in subsequent years, in spite of my own protests that I wanted them for the use of employees of my forestry department. It was but natural that the former owners of these various lands and buildings, living for many years in more or less penury, had exploited the very last sources of revenue connected with their possessions, especially the trees. Except in a few inaccessible spots, there was not a tree left on the entire Biltmore Estate that was fit for the lumber required by the Asheville market. Lumber in those days was very cheap. Even if there had been left a stand of yellow pine such as was found on the estate in the early days, how could I compete with the lumbermen of the East, who were throwing their lumber products on the local market at prices as low as twelve dollars a thousand feet board measure? Pinchot had bought a small portable sawmill, but he had abandoned it when he found that the expense of his diminutive logging operations was greater than the cost of better lumber bought and shipped to Biltmore from other sources.

And Biltmore Forest? From a German viewpoint, the forest might have been designated a chaos of trees belonging to a large number of species, many of them unknown to me. There were nine species of oak, of which black, Spanish, and post oak predominated. Chief representatives of the pines were a few fungus-ridden specimens of shortleaf pine (*Pinus echinata*). In addition, there were four species of hickory and remnants of chestnuts which, in earlier times, must have been the leading hardwoods in the stands. How did they come to an end? They had been killed by incessant forest fires set by the owners of the various farms to improve the pasture in the woods. The plowlands were restricted at that time to the bottoms along the French Broad and Swannanoa rivers, between which the major part of the Biltmore Estate was situated. What upland farms there were had been abandoned and were covered with sedge grass, a sight which, I was told, was an eyesore to Vanderbilt.

CHAPTER 4

Profile of Pisgah Forest

SOON AFTER MY ARRIVAL at Biltmore the American secretary of agriculture, J. Sterling Morton, arrived there, according to his promise, and asked me by note to meet him at three o'clock at Kenilworth Inn. I knocked at the door of his room at the appointed hour, and I was amazed to find him in his bed. What a sensation it would have been in Germany for a secretary of agriculture, not actually sick, to receive a foreigner in bed! Secretary Morton, however, received me most cordially, asked me a hundred questions, and later took me in a carriage on a trip of inspection of the estate. We saw the dairy and sheep farms, the beginnings of the pig farm, and the plan of the poultry farm. What forestry could I show him? None indeed! We might have inspected a few acres of abandoned fields which had been planted to white pines a few weeks before my arrival by an Illinois nursery firm. Unfortunately, these white pines were so small at the time, and so many of them had died, that an expert could not have found them.

Soon thereafter Pinchot came to Biltmore. Now, I hoped, I would get those insights and instructions for forestry at Biltmore which I could not frame myself. After all, forestry is common sense applied to woodlands. My common sense, unfortunately, was not sufficient for a forest policy on the Biltmore Estate. There was no forest nursery in which to raise planting stock; there were no markets for forest products and no byroads to make the woods accessible. The time for forest utilization either had passed or had not yet arrived. In the last analysis, forestry and lumbering, like mining and farming, are problems of transportation. The finest trees in the West in those days had little or no value because they were beyond the reach of railroads and steamboats. The situation in Biltmore, on a small scale, seemed to be identical with the conditions prevailing in the West on a large scale. There was one advantage for me, unique at the time in western North Carolina. A strong wire fence, six feet high, had been erected around that part of the woodlands situated on the right bank of the French Broad River. Thus it was easy for me to protect the major part of the Biltmore Forest from intruders, including fire fiends.

Curiously, on my trip with Pinchot over the Biltmore Estate, we continued to talk about hunting and fishing more than about forestry. These sporting discussions continued when we journeyed together on horseback to the Pisgah Forest, some one hundred thousand acres at the headwaters of the French Broad River which Vanderbilt had bought a few months earlier. The tract was some ten miles distant from Biltmore House at its nearest point, and some sixty miles away at its farthest end. On this memorable trip we took along our guns and some fishing tackle. Most certainly, no one meeting us on the road would have taken us for scientific foresters bent on serious work.

We spent the night at an inn, and the next morning we entered Pisgah Forest on a rough trail made by cattle driven by the natives to the pastures on the Pisgah Ridge. After a ride of some three miles, we hitched our horses on the top of the ridge and descended northward into the valley of Big Creek, in which, Pinchot had told me, logging operations on

a large scale were to begin. In the valley were the most beautiful trees I had ever seen — towering tulip trees, with gigantic
chestnuts, red oaks, basswoods, and ash trees at their feet.
Here, apparently, the fires had not played havoc. The Big
Creek Valley was a huge cove, and, as I learned later, the
survival of its glorious primeval forest had been due to its
inaccessibility.

Proceeding up the cove, Pinchot pointed out the site where
a splash dam was to be constructed. With its help the logs
were to be splashed into the Mills River, driven down the
French Broad River, and landed at the Vanderbilt sawmill
some forty miles distant from the cove. Apparently the plan
was ready: a survey had been made; the height of the splash
dam, twenty-two feet, had been determined; and the con-
tours of the water-storage reservoir above the dam had been
marked. Pinchot, apparently, knew all about it. He assured
me that Whitney, the ranger, on the basis of his experience
in the Adirondacks, was the proper man to build the splash
dam and to supervise all operations.

Splash dams? I had seen some splash dams in the Black
Forest, in Switzerland, and in Austria, all relatively small
and all used for the annual transportation of small logs from
the mountains to the lowlands. Here in Pisgah Forest the situation looked to me somewhat different. A splash dam was
to be built for one single operation; it was not to be a
permanent investment serving present and future forestry
operations. I was new to this sort of venture and, being a
German, I would have preferred the construction of permanent logging roads to the construction of a temporary splash
dam. Also, it seemed to me that road building was a prerequisite to permanent forestry. Roads were required, obviously, to bring into the heart of the cove the workmen, the
ox teams, the food for them, and the lumber needed for the
logging camps. Since roads were needed in any event, why
not build good roads of a permanent character? Furthermore,

no logs heavier than water could be moved by splashing and driving. This meant that no logs but those of the tulip trees could be floated to the sawmill at Asheville, and that the oak, ash, and chestnut trees would have to be left. I did not press these views, believing that Pinchot knew better; naturally, I accepted his superior knowledge.

After inspecting the site of our future forestry operations, we remounted our horses and descended by a meandering trail to the Pinkbeds, which were, and still are, a valley deriving its name from the pinkish color of the kalmia flowers (mountain laurel), whose bushes cover the bottom of its three thousand acres. In the Pinkbeds we found Lafayette Sorrels, a ranger who was a sort of police officer guarding the entire property against the usurpation of the natives living on some three hundred farms intermingled with the Vanderbilt property. The more fertile sections of the Pinkbeds were occupied by these farmers. It dawned upon me that the real owner of Pisgah Forest was not George W. Vanderbilt, but these mountaineers, who were using his property for farming, pasturing, and hunting at their own pleasure. In the proximity of the interior holdings and of the public roads, no trees of any value were left, especially in the huge valley of the Davidson River. Pinchot apparently took these inroads on the rights of the proprietor for granted. To my own European feeling, they were equal to theft and robbery.

Pinchot and I did not discuss these problems at the time. He was much more interested in the possibility of a bear hunt on the Pisgah Ridge, where a small log hut had been constructed for him and where, on the preceding day, upon his orders, a dead calf had been tied to a tree as bear bait. In the evening Pinchot and I climbed up to the hunting lodge with provisions in our knapsacks, guns on our backs, and high hopes of killing a bear. We watched alternately during the whole starry night. No bear came, nor did one make an appearance during the second or third nights. We found on the ridge, however, the carcass of a heifer that recently had been killed by a bear.

We stretched our tired limbs on improvised camp cots, but before doing so Pinchot knelt at his cotside in prayer. We had not discussed religion, but here verily was the most religious young man I had ever met — a man so accustomed to evening prayer that he did not abandon it on a hunting trip and in a dilapidated cabin. On a later visit, Pinchot delivered a lecture in the Biltmore church from the parson's pulpit. I did not understand a word of it, but I was told that, as a member of the Society of St. Andrew, he was required to bring back to the fold of saints annually at least one sinner. I have since wondered whether he expected me to be a prey for his endeavors! Of all my American friends and acquaintances, Gifford Pinchot was the most religious and the most abstinent. He never swore, drank, nor smoked.

The Olmsteds, my hosts and friends at Biltmore, had told me that Pinchot's reason for relinquishing his job at Biltmore was the death of his fiancée, Laura Houghteling, the daughter of a wealthy Chicago lumberman. The young lady had spent her dying months at a house on the French Broad River opposite the Biltmore Forest. The death of his fiancée was an explanation, no doubt, for the mourning clothes Pinchot wore at the time and for many years afterward.[1]

After a hunting and fishing trip of a week or so, Pinchot and I returned to Biltmore, where Vanderbilt had arrived in the meantime. He received us the next evening at a small dinner party at Brick Farm House, his temporary residence

[1] In 1893 Pinchot, with Vanderbilt's permission, had opened an office as consulting forester in New York City. After Schenck's arrival at Biltmore, the New York office served as Pinchot's headquarters until his appointment as chief of the Division of Forestry on July 1, 1898. During this time Pinchot served as secretary of the National Forest Commission and, with Henry S. Graves, wrote *The White Pine; a Study, with Tables of Volume and Yield* (New York, 1896) and *The Adirondack Spruce; a Study of the Forest in Ne-Ha-Sa-Ne Park, with Tables of Volume and Yield and a Working Plan for Conservative Lumbering* (New York, 1898).

while the huge mansion was being built. This first evening with him is unforgettable. Not for a moment was forestry discussed. Much more interesting to both Vanderbilt and Pinchot was modern French literature, of which both seemed to have a wide knowledge. My own small literary accomplishments at this meeting condemned me to silence. Vanderbilt was most kind to me personally, and he asked me to meet him at the Biltmore office the next morning at ten o'clock.

At that time Vanderbilt was a bachelor some thirty years old. He was a grandson of Commodore Cornelius Vanderbilt, founder of the dynasty, and a son of the late William H. Vanderbilt and his widow, nee Maria Louisa Kissam, a mother to whom George Vanderbilt seemed to be deeply attached. There were several older brothers, among them Cornelius and William K., as well as a sister, who had married Dr. W. Seward Webb, the owner of a large timbered tract in the Adirondacks. The sister and the brothers seemed to be little interested in their brother George's ventures in the wilds of North Carolina. Unless I am badly mistaken, they seldom came to inspect his Biltmore extravagances. Old Mrs. Vanderbilt, the mother, however, was a frequent guest for long periods when Biltmore House was completed.

How did George Vanderbilt look? Not like an American. He looked like a Frenchman, slender, with a black French moustache and with black eyes shining kindly as well as merrily, good humoredly, even jocosely. Unfortunately, this owner of a large country place did not care for fishing, hunting, golfing, riding, or driving in the country. His hobbies, nay, his vocations, were literature, landscape architecture, and interior decoration. When Biltmore House was completed, it contained forty-eight guest rooms exhibiting Vanderbilt's collection of furniture and equipment dating from early Renaissance through Louis XIV, Queen Anne, baroque, rococo, directoire, Empire, and what not, to the more modern styles of greater convenience. His collection of tapestries

Biltmore House in the Early 1900s

George W. Vanderbilt

Frederick L. Olmsted

Looking toward the Pinkbeds of Pisgah Forest and the Balsam Mountains

adorned the reception room, and his collection of the works of Albrecht Dürer was shown in a special print room. Most characteristic was his grand library occupying an entire wing of the edifice.

Vanderbilt had not seen any forestry — none in France, none in Switzerland, none in Germany, none in Sweden. He may have read about it, occasionally, in *Garden and Forest*. I do not know. Frederick Olmsted had spoken about it and so had Gifford Pinchot. He wanted my help as a forester for the management of his woodlands, and I was anxious to fill the job.

At our business meeting, punctually at ten o'clock A.M. in the Biltmore office, Vanderbilt was most kind. My annual salary was to be twenty-five hundred dollars. I would be provided with two saddle horses and their feed. For my quarters I would have the fine old country place heretofore occupied by Baron d'Allinge. I would be allowed a leave of absence for three months in 1896 and thereafter two months every second year in order that I might complete my military duties as a lieutenant in the German army.

Two of my requests, however, Vanderbilt refused to grant. When I asked that my salary be paid in gold, he explained that that would be impossible if William Jennings Bryan should be elected to the presidency in the fall of 1896. He also refused to supply the funds required for a good topographic map of the property, without which, I insisted, my task in Pisgah Forest could not be accomplished. A similar map for the 7,280 acres of Biltmore Forest had cost $30,000, and the expense for such a map, covering some 100,000 acres, he thought, would be out of proportion. At that, a sketch map already had been made showing the boundaries of Pisgah Forest and those of such interior holdings as were recognized as valid by Vanderbilt when he purchased the bulk of the property. That map ought to suffice for forestry, too, he declared.

I told him that forestry was a problem of transportation and that transportation was a problem of topography. I was enough of a surveyor to do the triangulation needed for a good topographic map, but it would take me several years, I said, to complete a job to which I could not give all my time. Vanderbilt remained adamant. Fortunately, McNamee, the estate manager, was well acquainted with the director of the United States Geological Survey and he knew that the Pisgah Quadrangle, among the maps of the Geological Survey, was about to be redrawn. For a nominal contribution, he thought, the map I wanted could be made. With this understanding, my contract of employment was signed.

I might interject here that the map desired was made in 1896–97 by the Geological Survey in co-operation with the state geologist of North Carolina. It was an excellent map on a scale of 1:45,000, with fifty-feet interval contour lines. It was the first map made in the United States for a large forestry undertaking. My own chief contribution to the map was the construction of more than two hundred miles of rough trails for the use of the surveyors, most of them along the ridges and all of them important in fire protection as well as in lumbering. The cost of trail building in those days was less than ten dollars a mile, a figure explained by the low wages then prevailing and by the fact that most of the trails had been prepared in a preliminary manner by pasturing cattle.

At my conference with Vanderbilt he told me, to my surprise, that Pinchot's connection with the Biltmore Estate had ended and that I was in no way subject to his orders or to his supervision. Vanderbilt thought it wise, however, for me to carry out Pinchot's logging plans on Big Creek in Pisgah Forest, inasmuch as the band-saw mill bought by him in Asheville and located on the bank of the French Broad River was waiting for logs to come from Pisgah Forest. As a consequence, Whitney was ordered to go ahead with the building of the splash dam on Big Creek in keeping with Pinchot's plans.

Evidently I was independent in all matters of forestry in Biltmore Forest as well as in Pisgah Forest, subject only to the financial control of McNamee, a condition that was most welcome to me on the basis of my German training. All payrolls, bills, salaries, taxes, and lawyers' fees were paid through him. My forest department never had a cent of Vanderbilt money in its hands.

CHAPTER 5

The Dam on Big Creek

I CONTINUED TO LIVE with the Olmsteds on the Biltmore Estate while my residence at "Woodcot" was being furnished. At that time another guest, the world-famous Charles Sprague Sargent, came to discuss with Frederick Olmsted the plan of the Biltmore Arboretum, to which I have already alluded. Sargent, of course, was deeply interested in the scheme, but he predicted failure unless Vanderbilt was willing to endow the enterprise from its start with a large sum — a million dollars, if I remember correctly. It was essential, Sargent thought, to free so long-going an enterprise from the whimsical fluctuations of personal predilections as well as from the fluctuations of personal finances.

On a visit to the Biltmore Nurseries Sargent surprised me when he distinguished from a distance of some twenty feet the several species of abies (silver firs) standing as seedlings two feet high, side by side, in a nursery bed. I myself was so ignorant that I would have failed to identify at close range the German silver fir and the North Carolinian balsam fir, stands of which I had seen on the Pisgah Ridge when hunting with Pinchot. But I was more than surprised when Sargent, overwhelmed by the gorgeous panorama of landscape viewed from the pergola at Biltmore House, asked me after a long pause in our conversation: "Dr. Schenck, why do men die while the tulip trees and the white oaks and the sequoias seem to live perpetually?" I did not know any answer at that time, and I do not know any answer to the question today.

A few weeks later, when I had moved into my residential "Woodcot," Sargent came again to Biltmore, this time as my own guest and accompanied by a man who looked almost as beardy, as unkempt, and as rough as any mountaineer in North Carolina. At lunch this queer individual made some bright and lyric remarks, causing me to exclaim, "Mr. Muir, you are a poet!" To prove the correctness of my bold assumption, John Muir sent me a little later an autographed copy of his book, *The Mountains of California*. Today there is no book in my library dearer to me.

In the meantime some hemlock logs had been snaked by cattle to the dam site on Big Creek and the foundation for the dam had been excavated. Work in Pisgah Forest began in earnest. For ten American dollars a native of Cruso, John Cogburn by name, built a little log cabin for me near the site, with a rough and very hard camp bed and a chair hewn from a cucumber log. A terribly poor and steep wagon road was made by Whitney across the steep ridge to the south of Big Creek, connecting the site with the nearest public road. What a fool I was! Why did I not insist that a really good road be built at an expense ten times larger but twenty times cheaper in the end!

My own chief job now was the study of similar operations of logging and splash dams in the southern Appalachians. Most interesting to me were those near Hickory, North Carolina, connected with a good band-saw mill there owned by some Northern boys who seemed not to have had too much experience of their own. They owned no woodlands, but bought logs delivered to their boom in the Catawba River. An unexpected flood in the river had broken their boom, and all their logs, most of them already paid for, had floated on to the Atlantic Ocean. The break was ascribed to the fact that the floating seam of the boom, which crossed the river diagonally and had been firmly attached in mid-river to some stone-filled cribs, did not rise freely with the river tides. The seam of their next boom, then building, was to be held fast to the riverbanks by wire cables, thus allowing the boom to rise and fall with the river levels. I concluded that their experience in boom building should be adapted for the boom to be constructed above our own band-saw mill in the French Broad River at Asheville, metropolis for Biltmore.

The logs required for the seam of my boom could not be obtained in Pisgah Forest. All of them had to be high quality logs cut from the best tulip trees, each log some forty feet long and from two to three feet through. A double row of such logs, their ends joined with heavy anchor chains, was to form the seam of the boom. Logs of these qualities were hard to find and expensive to buy along the lines of the Southern Railway, and it required many trips on horseback and much persuasion of their owners to obtain them.

On one of these trips I discovered a logging operation in Jackson County where a narrow-gauge railroad, a Shay locomotive, and a McGiffert log loader were supplying a small circular-saw mill with logs of all kinds and sizes. I was deeply impressed by the McGiffert loader. Here was a machine lifting itself by its own bootstraps and spotting the cars to be loaded through its own legs spread far apart! And this machine had been invented by a lawyer, John R. McGiffert. A few years later I made the personal acquaintance of this remarkable lawyer and got him to give some lectures at the Biltmore Forest School. Fifteen years later the whole Biltmore School visited him and the Clyde Iron Works in Duluth, for whom McGiffert was working. Great trees, great tulip trees, are interesting. So, too, are great men, and the greatest of great men are those who are unconscious of their greatness. McGiffert was one of them.

By way of contrast, let me introduce here another great man whom I met on my log-buying trips in western North Carolina. He was an old, age-worn chief of the Cherokee Indians living on a reservation. All the Indians except the chief lived in shabby cabins. The chief continued to live in a rough tent, from which he removed and showed to me, with great pride, scalps which he had taken in his early days, alas! from the blond heads of two white girls! The Indian boys of the reservation were exhibiting their skill by shooting small birds at distances of forty or more feet, with blowguns made of cane and arrows of stems of the cattail reed.

The Indians also showed me their forests, the most glorious stands of tulip trees with an undergrowth of tall beeches. I was eager to secure as many of the tulip trees as possible for my boom; but the Indians were the wards of the Great White Father in Washington and were forbidden to sell any of their property without his consent. A few years later a native lumberman obtained many of these trees by supplying the Indians with some liquor, saws, and axes. Thereupon trees were cut down and the Great White Father was forced to sell them for the benefit of his wards in order to prevent the cut logs from rotting in the woods.

With these experiences in dam building, some fine boom logs loaded on cars, and a genuine Indian blowgun, I returned to Biltmore just in time to welcome Fernow as a visitor. He was accompanied by a good-looking lumberman named Dan M. Riordan, said to have some large logging interests in Arizona and New Mexico and also some interest in American forestry. I was elated. Here was a chance to get actual advice for the affairs of Biltmore and Pisgah forests from two experienced men. But I was sorely disillusioned. When the visitors left after a short sojourn I was no wiser than I had been before. Neither Fernow nor Riordan had any practical experience in the construction of splash dams and booms. Fernow, in particular, was critical and faultfinding, where heretofore he had been admiring and appreciative of the Biltmore undertaking.

Upon Fernow's request I promised to receive as a forestry apprentice Graham M. Leupp, son of Fernow's friend, Francis E. Leupp, at the time editor of the *Washington Post*. Young Leupp was the first forestry apprentice among those who later became my beloved flock of Biltmore forestry students. Unfortunately, after a short stay with me and a longer stay in the German woods under the tutelage of Sir Dietrich Brandis, this promising youngster died early of tuberculosis.

The next great man to pay me a visit was Baron von Ribbentrop, successor in 1889 as inspector general of forestry in India to Dr. William Schlich. Ribbentrop, scion of a noble Hannoverian family, had gone to India in 1866 when the "English" kingdom of Hannover had been annexed by victorious Prussia. Unlike Brandis and Schlich, Ribbentrop was more a sportsman than a forester, and when he came to Biltmore he was eager to bag a large buck of Virginia deer. We journeyed together with our rifles to Pisgah Forest, where in the course of our hunting I hoped to obtain practical advice for my splash dam and for the logging operations about to begin on Big Creek. Ribbentrop was enough of a silviculturist to object to the removal, under Pinchot's plan, of all tulip trees excepting the hollow ones. Since we were cutting only tulip trees, whose progeny demanded much light from overhead for their development, our logging operations, Ribbentrop thought, were sure to extinguish forever our most valuable species of trees. Search as we might at that time, we could not find in the entire cove one seedling of a tulip tree, and not one tree of pole size.

Was I committing race murder among the tulip trees? When Ribbentrop left, what should I have done? Should I have stopped all operations? Should I have boycotted the Pinchot plan? Impossible! Pinchot was my friend and I was unwilling to betray him. Also, there was our band-saw mill in Asheville waiting impatiently for logs — two million board feet of them — required to make operations remunerative and to justify the construction of a boom; and Pinchot had promised Vanderbilt an annual dividend of four per cent or better when he advised him to buy the Pisgah Forest property. To make that promise good, all silviculture, all sustained forestry, all tulip trees, and my good name, if Ribbentrop was right, were to be sacrificed.

And so it happened that I threw Ribbentrop's advice and warning to the wind. Two months later, when Ribbentrop had seen Brandis in Europe and had explained my situation to him, Sir Dietrich gave me by letter his own opinion to the effect that, if the natural regeneration of the tulip trees should fail, I should plant them in the cove, deadening all old chestnuts and all old oaks that were in the way. Alas, Sir Dietrich overlooked the fact that there was a difference between the teak woods of Pegu and the primeval forests of North Carolina.

The splash dam, in the hands of Whitney, made no progress. Whitney, a Northerner, could not manage a crew of Southerners, especially a crew of Negroes. He was succeeded by Jack Malley, a man of gigantic strength who had had long experience in Appalachian logging and who possessed a still longer vocabulary of profanity. His authority with the crews, both black and white, was quickly established when he knocked down, with one blow, a Negro prize fighter unwilling to obey.

All cutting and snaking of logs, tulip trees only, to the creek bed were done under contracts drawn up by Charles McNamee, who was a good lawyer. Every log was seventeen feet long, top logs excepted, so that the logs might yield a large percentage of boards sixteen feet long. The bed of the creek was freed of protruding rocks and fallen timber and of all sharp bends that would obstruct the passage of logs driven in a splash. Where the creek had low and shallow banks, strong barricades were made along the banks to confine the current to a depth that would float the logs downstream. It soon developed, however, that our splash dam did not hold enough water to float the logs all the way down into the Mills River, a tributary of the French Broad. A second splash dam was constructed by Jack Malley on the North Fork of Big Creek, from which all tulip trees had been stolen or otherwise removed many years before Vanderbilt purchased Pisgah Forest. The waves from the two splash dams were so timed that they met one another at the confluence of the two creeks. Unfortunately, both dams were high and the seepage of water beneath them was considerable, with the result that it usually took two weeks after a splash to refill the reservoirs above the dams — two weeks during which there was no work for the crews, while the payrolls for waiting labor were mounting.

The water in the Mills River, a public stream outside Vanderbilt's property, was not high enough, ordinarily, to float the large tulip tree logs. Thus we were forced to wait for rains and freshets to augment the flow. When they came, my worries were hugely increased. Logs were spilled over the fields framing the Mills River; riverbanks and bridges were torn; farmers along the river were furious, and my crews were forbidden to enter upon their premises, where many logs were stranded and scattered. A very rattail of lawsuits was the consequence. Malley, by intimidating the farmers, proved to be a most valuable asset. In the end, Vanderbilt was compelled either to pay the bills for damages presented by the farmers or to go without the logs. Had he seen the terrible change presented by Big Creek, the devastation in the laurels framing its once paradisiac banks, the rocks in the creek washed bare of their original mossy patina, he would have been furious, and forestry in Pisgah Forest might have come to quick death. Fortunately, he was absent, spending the fall and the winter in Paris.

Meanwhile Pinchot had been busy with a plan, first promul-
gated in *Garden and Forest* in 1889 by Professor Charles S.
Sargent, to have President Cleveland appoint a forestry commission to study and prepare for Congress a report on the
administration and control of the western public forest lands.
In addition, Pinchot was writing a book on the white pine and
developing a plan for forestry on the Webb Estate in the
Adirondacks. No wonder that he had no time for me and for
my worries! He had jumped suddenly into the limelight of
public importance.

My own troubles were lightened somewhat by help that
came to me accidentally. Some twenty miles from Biltmore
lived the Westfeldt family at Rugby Grange, one of the better-
kept, old-fashioned Southern estates that had survived the
Civil War. Young Overton Price was living there, his mother
being a Westfeldt. He wanted to become a forester and asked
me to allow him to work with me as an apprentice. He soon
developed into a good helpmate. He knew the country and the
language; he knew the native trees; and he was a capital man
on horseback. He was soon joined by another apprentice, E. M.
Griffith, a college graduate like Price, and like Price indefatigable in helping me to measure cordwood, to secure needed
supplies, to correct my English, to survey auxiliary roads in
the Biltmore Forest, and so on. I was very lucky also in my
office force, which consisted of one woman, some thirty years
old, who was forced to support her four children and her
husband by working as a stenographer and bookkeeper. Her
education had been excellent, and the letters I dictated to her
in bad English came out of her writing machine immaculate
and perfect. I shall be grateful to Mrs. Eleanore G. Ketchum
to my dying day for the knowledge of the English idiom which
she inculcated in me.

Price and Griffith worked for me without any remuneration.
Remuneration! Naturally, forestry must be remunerative if it
is to exist anywhere on a large scale. To that end, the products

of forestry, the trees, must command a price sufficiently high to make their production remunerative, and the future of tree investments must look bright and must be reasonably free from reckless competition, forest fires, and other hazards. None of these conditions prevailed in the United States in 1895; and yet I felt that it was my mission to prove to the United States the possibilities of forestry as a long-term investment. I soon realized that German forestry, the variety in which I had grown up, was as impossible of success in the United States as was Indian or Swedish forestry. A brand-new sort of forestry was needed, varying from state to state, and from county to county. It now dawned on me that, if it was impossible for me to practice forestry on American soil in 1895, I might at least lay the foundations for its practice at Biltmore in 1945.

In the Biltmore Forest my next problem was the disposition of those thousands of cords of wood, mostly from dead chestnuts, which Pinchot had left scattered over the forest. Again, "forestry is a problem of transportation." In Switzerland, I had seen portable chutes, troughs made of boards through which small logs and fuel wood were shot toward the valleys by gravity from mountain tops and from steep slopes. Pinchot's wood was moved by my teams over rough roads, consisting merely of ruts to hold the upper wheels to the mountain spurs pointing toward the French Broad River; and chutes were set up on these spurs, one after the other until all wood relics had been shot to the river and floated into the boom then under construction at the band-saw mill. There the wood was cut up in keeping with the desires of customers with the help of a steam-driven swing saw and a splitting machine. Alas, the good public must pay for whatever it gets: good houses, food, comfort, transportation, and — last but not least — good forestry!

CHAPTER 6

The Cloudy Dawn of
Private Forestry

ON ONE OF OUR HORSEBACK RIDES over the Biltmore Estate,
Vanderbilt had shown me the "eyesores" of abandoned slopes
covered with sedge grasses, which he wanted to have planted
in trees with the utmost speed. In some of these old fields a
few yellow pines, seeded in over long distances, were beginning to show themselves. But Vanderbilt did not want yellow
pines. He wanted to re-establish the same hardwood forest
that had theretofore prevailed. It seemed to me easy to produce the stands he desired. All I needed were a few hundred
bushels of acorns and hickory nuts, which could be bought
cheaply. They were obtained and planted, five bushels to the
acre, in shallow furrows made with a bull-tongue plow, and
covered with the dirt of another furrow.[1]

A planting of hardwood seedlings made by Pinchot close
to the Biltmore mansion in the early spring of 1895 had been
a failure, although the tulip trees, the walnuts, the basswoods,
and the maples had been planted with the utmost care on
well-prepared ground. Eighty per cent of the planted trees,
knee high, had died from lack of moisture. The moribunds
were cut back to the ground with pruning shears, and a majority of them were saved when the pruned specimens formed

[1] The results are described on page 55, below.

45

new stump sprouts. This method of afforestation, it seemed to me, was insecure and more expensive than the planting of nuts and acorns.

Within the Biltmore woods no seedlings were ever planted. Here I relied on natural seed regeneration, helped in the case of the oaks and the hickories by student workers pushing the nuts lying on the ground into the soil with the heels of their shoes or with ramrods after heavy fall rains. Where seedlings of yellow pine had developed and were overshaded by misshapen hardwoods, the latter were cut and made into fuel wood. Misshapen stump sprouts of the oaks, the result of repeated fires in the past, were also cut off, and, lo and behold! the new sprouts obtained after cutting were healthy and promising!

I thought it essential to press into my memory the looks of the woods as they had been before improvement cuttings and as they developed in the course of the years following such cuttings. Some twenty-four places were selected and marked permanently, and annually, from 1895 on, they were photographed by an expert with the same camera in the very same direction, at the very same hour and day of the year. By these means I obtained, at a time when movies were unknown, twenty-four sets of pictures showing the effect of my ax work and the development of the woods under my forestry. The pictures are almost as instructive as if the scenes had been filmed by a motion picture camera.

Vanderbilt, fortunately, approved of my infatuation for building roads on the Biltmore Estate, especially roads which he and his guests could use as riding roads and which at the same time facilitated the transportation of fuel wood to town. A service road also was built and macadamized, saving distance as well as grade between some three thousand acres of woodlands and the city. Every tree cut in the woods was marked by me personally. Before a compartment for an improvement cutting was marked in, rough roads were laid out on a downhill grade connecting with the standard dirt roads and through them with the macadam roads. Cut wood was piled by the woodcutter along the nearest rough road. All woods work was done by contract and all wood cut was fuel wood for the Asheville market. No wood was cut beyond the reach of roads and teams. Why bring up these details? Many a farmer in western North Carolina has come, I am told, to imitate this scheme of forestry and roads in his own small woodland; and many a farmer in New England, in the Lakes States, and in the West might follow the example to his advantage. In our forestry work everywhere, the construction of roads of the first, second, and third orders is essential.

In the meantime, great things had happened in Washington, D.C. The National Academy of Sciences, at the request of President Cleveland, had named seven Americans to form a National Forest Commission to report on the future of that part of the public domain in the West that consisted of forest land unfit for farming. The commission was headed by Charles S. Sargent; its secretary, selected by Sargent, was Pinchot. I was surprised at the time that leaders in American forestry such as Fernow and Rothrock were not on the commission; nor was any native-born Westerner.

Pinchot, accompanied by Henry S. Graves, who had just returned from studying forestry abroad, had the wisdom to visit the northern Rockies on a study trip of his own in the early summer of 1896, several weeks ahead of the commission. After touring the West from Montana to Washington and from Washington to New Mexico, the commission submitted a report on May 1, 1897, recommending the creation of thirteen forest reserves, comprising some twenty-one million acres, to be protected by the United States Army. Here the example of the French and that of the Indian Forest Service seems to have served for a precedent. Obviously, the time for forest activities in America had not arrived. Forest protection was needed first.

Sargent as president of the commission and Pinchot as its secretary seem to have disagreed on many points, among them the provision for military supervision of the reserves, which Sargent advocated and Pinchot opposed. This fact, however, did not prevent Pinchot from signing the commission report. By a proclamation issued by President Cleveland on February 22, 1897, the reserves proposed by the commission were created. There followed a Congressional hullabaloo in Washington, the Senate voting to void the proclamation and the House, to support it. The differences between the two houses ended in a compromise, which apparently was not to the liking of President Cleveland, who refused to approve the bill. Sargent was sorely disappointed by his differences with Pinchot and other members of the commission, and in the course of the year he abandoned forestry and his publication of *Garden and Forest*. Pinchot, on the contrary, remained sanguine.

Letters written to me at the time by Sargent and Pinchot indicate their feelings. Sargent wrote on April 14, 1897, from the Arnold Arboretum:

Of course we shall be more than glad to publish in Garden and Forest your paper on forestry and I trust you will send it to me at once.

I do not feel very sanguine about the outcome of the reservations and I am afraid that we are never going to have any forestry to be fathered by me or by any one else. There are horrible political complications in this matter which I won't bother you with today.

I am very sorry to hear that things are not going entirely to your liking in the Biltmore forests, although I confess I am not surprised. The time has hardly come yet in America when private forests can be managed systematically. I should like to know just what Mr. Pinchot's relations are now with Biltmore. Has he any supervision or control over you in any way? Of course anything you may say on this subject I shall regard as strictly confidential.

I am not without hope of a visit from you here, but I trust

Biltmore Students and Their First Schoolhouse

The Splash Dam on Big Creek

Sawyers at Work in Big Creek Valley

Logs in Big Creek Awaiting Splashing

that will not mean that you are leaving America where your services are so much needed. . . .

And from New York on April 19, 1897, Pinchot wrote that he hoped "for much good, instead of harm, from the attack on the reserves," and that it would "not be unexpected" if the nucleus of a forest service were obtained the following summer.

Obviously, Pinchot was right and Sargent, uncompromising as he was, was wrong. It was easier, no doubt, for Pinchot to yield than for Sargent, who must have felt that his reputation as the leader of American forestry had been shattered.

In the meantime my own dissatisfaction with forestry in Biltmore had led me as far, almost, as it had Sargent. I was discouraged and on the verge of quitting. When my logs rescued from the Mills River arrived in the French Broad, that river did not have water enough for six months to float the logs to the mill in Asheville; and when the rains and the freshets came, they were so wild that many logs dived beneath or jumped the boom and were routed for the Tennessee River and on to the Mississippi! Adding to my worries, McNamee insisted that all sales of lumber should go through his office and, without consulting me, he had appointed an insurance broker living obscurely in New York City as general lumber sales agent. This man was eager to earn his commissions quickly and, to facilitate that end, he accepted any price cut which a customer in New York requested on the arrival of a car of my lumber.

Pinchot again came to Biltmore, and together we visited the site of his and my logging operations in the Big Creek Valley of Pisgah Forest. He was highly pleased. Wherever the soil had been scratched by the logging operations, there was a matting of seedlings of young tulip trees. Evidently the old majestic tulip trees felled by us were loaded with seeds at the time of cutting and they had immortalized themselves by the profusion of their progeny!

What is forestry? If forestry consists merely of the replacement of a fine stand of primeval trees by an abundance of seedlings, Pinchot's plan of forestry had been gloriously successful. Financially, it had been a debacle; but, silviculturally, it seemed to be a great success. I confess that I myself, striking a balance as best I could, felt guilty. Unique, unrivaled wonder trees had gone, and had gone forever. The primeval beauty of Big Creek had been destroyed. The fertility of the cove had been reduced by the acceleration of its drainage. The financial loss incurred by our brand of forestry amounted to many thousands of dollars.

Yes, Pinchot was elated, while I was utterly depressed and torn by a doubting conscience. Was he correct? Can we apply the term "silviculture" to a process by which the finest trees of the woods are replaced by seedlings of doubtful prospects? Is a war won when all men and soldiers are killed, to be survived merely by a multitude of their children? I do not know. Nor do I know what happened to those seedlings later on — whether they have been killed by raging fires, by suppressing oaks, by fungi or insects, or by cattle. Many of them had developed and were saplings when I left Pisgah Forest in 1909.[2]

Whatever the case may be today, Pinchot as the originator of the logging operations on Big Creek and I as his successor cannot escape censure. We acted without due consideration of means and ends, lacking a knowledge of local conditions, of lumbering, of markets, of freight rates; and we disregarded the financial as well as the forestry interests of our employer. If seedlings of yellow poplar were wanted, we might have obtained millions of them by scratching the soil, by removing the leaf litter on the ground, by exposing the mineral soil for the reception of yellow poplar seeds. A glance at some of the abandoned fields in the coves of Pisgah Forest, notably

[2] The Big Creek area was logged again about 1930 by the Carr Lumber Company.

those on the north slope of Avery's Creek, would have taught us the lesson. In other words, at a small expense, and without sacrificing the finest trees, without destroying the gorgeous beauty of the woods, without incurring a financial loss, we might have obtained the same end — seedlings, a billion seedlings of tulip trees. This visit of Pinchot was his last to Pisgah Forest. My friendship and my admiration for him continued unabated. His conception of forestry differed from mine, but a friendship that cannot stand an occasional difference of opinion does not deserve the name.

At that time the number of forestry apprentices at Biltmore had increased, and on rainy days I began to give them some talks, or "lectures," on forestry, all fresh from my European textbooks, notably from Schlich's *Manual of Forestry*. There were many discussions in the woods and in the office, which served as a schoolroom. I learned on these occasions from my "students" much more, I am sure, than they learned from me. The apprentices were invaluable in nursery work, cone collecting, and seed extracting, and in constructing trout ponds and a fish hatchery; and they acted as quick and reliable messengers between headquarters and foremen and forest rangers, whose numbers had risen rapidly to seven, four of them on the seventy-five hundred acres of Biltmore Forest and three of them on the hundred thousand acres of Pisgah Forest.[3] Every ranger was required at the week end to submit with his payroll sheet short reports on fish and game, roads, nurseries and plantations, logging, wood cutting, forest fires, and trespasses.

A large part of my time during 1895 and 1896 was spent on horseback trips in Pisgah Forest. I wanted to become acquainted with the boundary lines, exterior and interior, and to see the woods and their possibilities of utilization. Alas and alack! Those possibilities were meager. Yellow poplars were

---

[3] The acreage of the Biltmore Estate and Pisgah Forest was constantly changing as land was bought or sold to consolidate the Vanderbilt holdings.

scarce, and only yellow poplar had any value. There was no further possibility of driving logs on any of the streams draining Pisgah Forest because all river valleys were settled. There was no possibility of logging to the public roads because it would have been necessary for our logging roads to traverse these settlements. In the southernmost part of Pisgah Forest the size and the number of the interior holdings were so great that Vanderbilt's property in the aggregate was smaller than that of the holders. The woods in my charge were on the ridges and on the slopes above the farms where there was no yellow poplar. Mine seemed a hopeless task. For years to come, I could not think of conservative forestry.

It dawned on me that the elimination of the interior holdings by purchase, and thereupon the abolition of the public roads and their conversion into private forest roads, was a prerequisite of any prospective forestry. At that time Vanderbilt was offered for purchase a tract of some ten thousand acres on the Pigeon River northwest of the Pisgah Ridge at a price of one dollar an acre. I looked over the tract. It was heavily stocked with spruce (*Picea rubens*) and fir (*Abies Fraseri*), and it had not suffered from fires. There were no squatters nor interior holdings in it. Vanderbilt was inclined to purchase it; but I, fool that I was, dissuaded him on the grounds that the tract was inaccessible from Pisgah Forest. The acquisition of the interior holdings within Pisgah Forest seemed to me to be far more urgent and advisable than the purchase of outside lands. Ten years later the same tract was bought by the Champion Fibre Company of Canton, North Carolina, for ten dollars an acre.

I was blind in those days to the financial possibilities and to the rapidity of the economic changes going on in western North Carolina. Verily, a good forester, working for the future as all foresters do, must look ahead. But where is the German forester who has ever correctly foreseen what the future had in store? And developments by leaps and bounds, such as have taken place in the United States since 1895, are unique not only in the history of forestry but in the history of the world.

The lumbermen of the United States knew better. They had purchased even in those days all available tracts of long-leaf and shortleaf pine in the Southern states in anticipation of the exhaustion of the white pine in the North. One of the best of them, Frederick Weyerhaeuser, had the foresight to secure immense holdings in the far and undeveloped West. I well remember that the lumber papers at the time were aghast at the audacity of Weyerhaeuser's purchases.

Frederick Weyerhaeuser! His cradle had been in the hamlet of Niedersaulheim, some fifty miles distant from Darmstadt, Germany, my own native place. He had had none of the educational advantages that I had had. He landed in the United States as a poor immigrant, and, unlike myself, he did not arrive in a first class cabin of the best ship of the North German Lloyd Line. But he had two qualities that I lacked. He wanted to make money and he was qualified to make it; and he had vision. I did not want to make money — I wanted to practice forestry; and I lacked vision. Yet if I were to live my life over again, I would not wish to live it differently.

In the meantime the logging operations and the splashings on Big Creek proceeded at full speed. O those splashings! The bed of Big Creek, arched with rhododendrons, green with moss-covered rocks, replete with brook trout, was made a ruined run, a veritable arroyo of torn shores and skimmed stones. Is that American forestry, real forestry, conservative forestry? In Germany, too, in the Black Forest on the waters of the Murg River, the same vandalism had been perpetrated a hundred and fifty years earlier under the label "forestry"! Indeed, where has it not been true that forestry, superimposed upon the primeval, has destroyed nature's gorgeous beauty at the first stroke of the ax? In this respect the lumberman is scarcely more criminal than is the forester. Forester and lumberman differ in the last analysis, but in this, the lumberman destroys the primeval woods and the primeval regimen of the waters for business' sake; the forester does it for a chimera.

But here I ask, is forestry itself not a business? If it is not, it has no room in a country as businesslike as is the United States. And where in the world of the woods is lumbering not a legitimate part of forestry? The infantile maladies of American forestry are due largely to the fact that, in public opinion and by the agitation of foresters led by Pinchot, a strict contrast was established between forestry and lumbering. Nowhere else in the world does it exist.

CHAPTER 7

The Germ
of a Forest School

IN THE SPRING OF 1896 I was eager to see sprouting the acorns and the nuts I had planted — so many hundreds of bushels of them — in the abandoned fields of the estate. My plantations were complete failures. Frederick Olmsted, Jr., and I, on our hands and knees, tried hard to find some seedlings hiding in the sedge grass. I had not figured on the field mice in the case of acorns, nor on the squirrels in the case of nuts. Most of the seeds had been eaten by the rodents; the few that had escaped were hopelessly smothered by the dense matting of sedge grass, the roots of which were in firm possession of the ground. I was sorely disappointed and ashamed. Vanderbilt wanted those fields planted immediately — "if possible sooner than immediately" — a phrase often in the mouth of Sir Dietrich Brandis. But Vanderbilt did not have a word of blame for me when he learned of the failure. He was kind enough to realize that I had tried an experiment; and he must have been confident that I would learn by my own failures.

What could I do to redeem the failure? I had no seeds to sow in a nursery from which plants might be raised. There were no American commercial nurseries from which plants might be bought. In despair I wrote to some commercial nurseries at Halstenbek, near Hamburg in Germany, asking

55

if they thought it possible to ship across the ocean to Biltmore tree seedlings for my planting. The Halstenbekians were delighted to try the experiment, and about a hundred white pine seedlings were sent at once, as samples and by mail. They were planted on an old wood road in Compartment 106 of the estate and they all lived. Thus encouraged, I placed an order in the spring of 1897 for five hundred thousand white pines — two-year seedlings and two-year transplants — white pines being the only American plants available in the German nurseries. The plants arrived in New York and were reshipped as fast as possible to Biltmore, where they arrived after a seven-weeks voyage.

The majority of the shipping containers were willow baskets some forty inches high and thirty inches wide. The plants, in loose bundles, were packed in moss, the roots in the center, the tops near the circumference of the basket. They were in excellent condition. I had these trees planted south of Compartment 80 — the Apiary Plantation, named from an abandoned apiary found there — and between Compartments 17 and 18 — the Ferry Farm-road Plantation — on the most abandoned fields of the estate. To prevent the seedlings from being torn loose by the storms on these bare and wind-swept fields, I had stones placed on the roots of most of the plants. The success was unique. Vanderbilt was highly pleased. The purchases of white pine seedlings in Halstenbek were continued until my own forest nurseries were ready to supply the plants required for afforestation.

Most of the abandoned fields afforested were traversed by deep gullies. Here erosion had to be combated, as in the Alps, at the points of origin. At the upper ends of the gullies rough wickerworks a few inches high were made in semicircles or in parallel rows. In the gullies stronger barricades were made of poles deeply set and of branches interwoven or laid from pole to pole. The scheme worked marvelously. Within the short period of five years the gullies disappeared and within ten years the worn-out fields were covered by thickets of white pine.

The seedlings of the native broad-leaved trees — walnut, sugar maple, cherry, etc. — which I planted on the old Biltmore fields did not develop satisfactorily. The walnuts grew crooked, the maples suffered from spring frosts, the black cherry grew bushy. On a few fields, thinking the method to be cheap, I tried planting seeds of the black locust in parallel furrows, but with poor results. Here and there, at fairly regular intervals and apparently on the very spots where a heap of manure had been lying for days past, there were small groups of seedlings and, later on, of saplings; nothing between them. While it seemed to thrive in waste places, black locust I found to be an exacting species. I did not know it as such. My luck might have been better if I had planted the eroding fields with seeds or seedlings of sassafras.

The trees that I used on the abandoned fields in Biltmore with success in later years were for the most part native pines — notably shortleaf (*Pinus echinata*) — as visitors to the estate now know. These native pine seedlings were raised in my own nurseries, which, unfortunately, had to be situated on poor soil because the farm department monopolized all fields likely for crop production, and there was no surplus of them. But I could obtain plenty of manure from my own stables, and I enjoyed the astonishment of onlookers when I mixed the manure with the soil of the nurseries glovelessly with my own fingers. There was a barn, too, which I used as a seed-extracting plant; unfortunately, it burned twice when the heating stove had been carelessly attended by the ranger in charge.

Under my contract of employment I was allowed, for the first time in 1896 and thereafter every two years, a leave of two months to comply with my military obligations in Germany. From my visit in 1896 I returned a married man. My fiancée, Adele Bopp, had remained true to me and had the courage to accompany me to a faraway land whose language she did not know. On the trip across the water, in the "Trave" of the North German Lloyd Line, we were accompanied by German servants named Spiess, who also were newly wed. He was an ex-dragoon of the army, and she a cook highly recommended. In addition to an enormous amount of baggage — six huge trunks and several boxes containing household goods for our cottage at Biltmore — we carried with us two German dachshunds. No wonder that the fellow passengers on the "Trave" suspected that we were a princely couple in disguise; and my young wife was indeed lovely enough to emulate any princess of royal blood. Upon arriving in New York our two dachshunds were objects of great curiosity. We were frequently asked what sort of animals they were. We did not stay in New York, but took the night through train to Biltmore, where we arrived with servants and dachshunds on the afternoon of August 12, 1896.

Pardon me, dear reader, if I seem to mix in these pages family affairs with those of American forestry. Mrs. Schenck, however, has a good right to occupy a few lines in the annals of American forestry. In the early days of forestry she was hostess for many American and foreign foresters. At the Biltmore Forest School, when it was established later, she was a motherly friend of "my boys"; she knew and she liked the mountaineers. She was a model wife for an American forester roughing it in the backwoods. Often did she sigh and smile in later years: "I can well understand why a young man takes forestry for his life's career; but woe to the young lady who takes a forester for her life's husband."

My work in Pisgah Forest became increasingly diversified. Vanderbilt had resolved to build a huge hunting lodge at the Buckspring, close to the summit of Mount Pisgah. The walls were to consist of chestnut logs, and about a thousand logs of that species, ten feet in diameter and up to forty feet long, had to be supplied. Unfortunately, logs of that descrip-

THE GERM OF A FOREST SCHOOL

tion do not grow on mountains five thousand feet high. Therefore a wagon road some four miles long on a grade nowhere exceeding six per cent had to be built from the valley west of Pisgah and from the hamlet of Cruso in Haywood County to the site of the lodge. With the help of my forestry apprentices, whose number had steadily increased, and with the aid of that capital clinometer, the Hessian Bosé, the road was built in a few weeks. For the thousand logs, contracts were let with the farmers in the Pigeon River Valley, and when it was found impossible to obtain the chestnut logs required, the architects consented to accept logs of any species, provided they were of the right sizes.

I did not like the idea of the great hunting lodge. I would have much preferred to erect some fair houses for the increasing number of my rangers and for the workmen permanently employed. But Vanderbilt was not so inclined, not realizing that a good employee must live with his family contentedly in a decent house. As a consequence, most of the men whom I tried as rangers left their jobs when they began to know their districts and their duties. Vanderbilt was opposed also to my proposition to have the common workers, white or black, live on the Biltmore Estate. On the contrary, he required the houses on the many holdings comprising the seventy-five hundred acres of the estate to be torn down as soon as they were acquired. Never could I prevail on him to evince any interest in my rangers or their families, with the result that my endeavors to attach the ranger families to the Vanderbilt cause were frustrated.

At my house in Biltmore I established a small deer park of my own to get acquainted with the feeding and breeding habits of the native Virginia deer. The deer increased in their narrow enclosure so rapidly that some of them had to be turned loose in Pisgah Forest from time to time. On one occasion, when Gifford Pinchot entered the fenced parklet against my warning during the rutting season, a big buck,

Monarch by name, offered fight. Pinchot might have been hurt had I not lassoed the buck's forelegs, pulling him down to the ground. And on another occasion a young buck, after being liberated, attacked a girl, throwing her down and hurting her badly. Here I found that tame and semitame bucks are more dangerous than are wild ones.

A cousin of Vanderbilt's, with whom I had been quail shooting on the estate, went to him with a plan for a pheasantry on the English style. I bought some Mongolian pheasants, and George Gillespie, a ranger, was placed in charge of the pheasant breeding. Unfortunately, the pheasants proved to be vagabonds and fell prey to the guns of the farmers living in the proximity of the estate. After a few years the scheme had to be abandoned because I could not persuade our neighbors to spare the birds liberated until they had increased in the open woods by themselves and had stocked the entire country.

In a small and cold brook on the Biltmore Estate a trout hatchery was placed in charge of Cyrus T. Rankin — the best and most loyal ranger I ever had. There were six ponds, the water falling from one pond into another; and there was a hut in which the trout eggs obtained from the United States Fish Commission were hatched with the greatest possible care. The fish developed well, but in 1899 when there was an unprecedented cloudburst and the little brook feeding the ponds became a raging torrent, the ponds were completely destroyed and all the fish were lost. Worse than that, a beautiful small valley was converted into a series of ugly gullies. Failure after failure! Disappointment after disappointment!

At the village of Biltmore I had built a small steam plant, patterned after one I had seen in the Sihlwald near Zurich, in which fuel wood cut on the Biltmore Estate was dissected and split in such sizes as the Asheville customer wanted for his stove or open fireplace. This woodcutting plant gave me an advantage over the farmers who sold fuel wood, always in eight-foot lengths, on their trips to town. In addition, unlike the farmers, I could make quick delivery and furnish hickory only, or oak or black gum, for backlogs in the open fireplaces. I could ask a much better price than my competitors could obtain. When the system of macadamized roads on the estate had been completed, it was possible to cut and haul fuel wood every winter from any section of the estate, via the wood-splitting plant, to customers in Asheville. This was done by contractors as well as by the teams of the forest department of the Biltmore Estate. I had bought a dozen or so mule teams from the landscape department when it was reduced in size. Every year some three thousand cords of fuel wood were cut and sold at a small but distinct profit. I did not mark for cutting any tree that was not decrepit or inferior to its neighbor. And never, during my entire time of residence, was a tree cut on the estate which I had not marked personally for removal.

In my marking, my forestry apprentices were invaluable aids, free of cost to the employer. They accompanied me everywhere and they asked continually for explanations: "Why do you do this? Why do you cut this tree and not the tree yonder? How many seedlings do you figure per acre on this tract to be afforested, and how many on another site? What is the freight on lumber to New York and what is it to Cincinnati? How is the lumber market this week? What do you understand by 'forestry'?" From these and a thousand other questions, the idea of the Biltmore Forest School originated two years before it opened officially.

My youthful companions were teachers for me also; they taught me to speak English without fear of making too many linguistic mistakes. Dr. Fernow, whose wife was American born, had a really astounding command of the English language, and was in a much better position than I to use it. About that time Fernow was trying to consolidate the forestry movement in the United States by forming a union of the various state forestry associations. There was no such association in North Carolina, and he asked me to assist him in forming one at a meeting in New Bern in March, 1898, and to deliver an address on any topic I might select. I concocted a lecture on "Our Commonwealth and the Necessity of Forest Preservation," and, being timid, I had it corrected and printed, intending to hand it to the audience in print in lieu of an oral delivery. Mrs. Schenck had placed in my traveling bag a bottle of red claret which I had secured at Tryon, North Carolina, from a Swiss friend of the Swiss landscape foreman, old Frederick Brandley. En route to New Bern the claret inspected my printed address, soiling the copies with red spots.

At the meeting Fernow delivered an opening address with his usual vigor and eloquence. The audience was enthusiastic for forestry. I was next on the program; but I refused to talk, explaining that I did not have sufficient command of English and that my printed address, intended to take the place of my talk, had met with a fatal accident in transit to the meeting. I exhibited my soiled sheets. The audience smiled, then laughed heartily. Encouraged by this good humor, I babbled along as best I could for a few minutes. When I ended, there was much applause, and Fernow, embracing me, exclaimed: "Dr. Schenck, you are a born orator." I have been thankful to him ever since for that compliment. From that moment on, I have never had fear of speaking in public. Dr. Fernow had rid me of that fear.

CHAPTER 8

The Pinkbeds
and the Mountaineers

THE ENTIRE SUMMER OF 1897 I spent in the Pinkbeds of Pisgah Forest, where I had taken possession of the Sorrels' ranger house for household and office. Near by was a capital spring covered by rhododendrons. Here I really began to get acquainted with the mountaineers, their little lives and attitudes. There were two groups of these mountaineers.

In the first group was Mrs. Jerusha Owenbey, a kind old lady living in a two-story house built of lumber. Whenever there was sickness anywhere, Mrs. Owenbey was the first and last to administer aids; to watch whole nights at a sickbed; to apply homemade medicaments; and to manage the household of the afflicted if the mother was ailing.

In this group also were Artillus and Rufus Rucker, men about twenty years of age, thirsting for religion, for divine service on a Sunday, and for singing church anthems used by the Methodists and the Baptists. There was the rich man of the settlement, Hiram King, a carpenter by profession, who had built, with archaic means, a sawmill driven by a flutter wheel on Lookingglass Creek, and who had a number of beehives. King occupied the largest and best house in the Pinkbeds and he owned a small farm.

There was Ananias McCall, a kind soul who owned a mis-

erable starvation farm and an equally miserable log cabin at
the head of Avery's Creek, near Chubb, alias Club, Gap. He
had been baptized "Ananias" because his father, looking for
a name for the newborn babe, chose the name which he struck
first with a pin when he opened his Bible.

On the north fork of Big Creek, close to Beaverdam Gap,
lived Ulysses Reeves with his widower father. Ulysses was the
strongest man of the settlement, who, lacking a mule, carried
a hundredweight of flour from the Melvin N. Stuart mill to
his home over four miles of rough roads and trails. Ulysses
had ambition. He went to school somewhere and became my
most learned, but not my most faithful, forest ranger.

The other group lived by moonshining instead of farming.
They had their liquor distilleries in the mountain coves,
and shifted them from site to site to avoid discovery. They
went about armed, keeping the others in awe and threaten-
ing death to any betrayer of their secrets. To this class belonged a family on upper Davidson River. The mother was
in the habit of washing the family's soiled dishes but once a
week. Passing by one morning, I found two young men at
the cabin, cleaning it out, airing the bedding, mending shoes,
and removing the refuse accumulated outside.

"Where do you come from? What are you doing this for?"
I asked. "Well," was the answer, "we are Mormons from
Utah, and we are pledged to move about and to do good as
best we can." Mormons! I had heard of them and that they
were bigamists. And here they were, acting like the best of
Christians.

Viewing these mountaineers, I was in a real quandary. If
I were to supply them with a schoolma'am and a small organ
for the church schoolhouse, if I were to erect a grocery store
and help to improve their farms and living conditions, I
could never hope to get control of the interior holdings, the
acquisition of which was essential, I thought, for my forestry
plans. On the other hand, I was anxious to have the best ele-

The Schenck Summer Home in the Pinkbeds

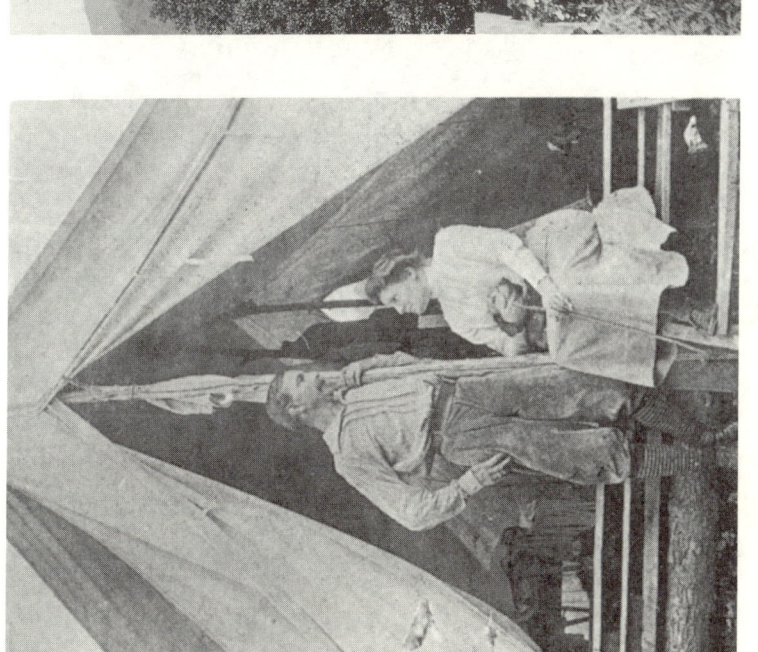

Dr. and Mrs. Schenck in Pisgah Forest

Lookingglass Rock

ments with me in the woods to supply the needed labor, to
assist in protecting the forests from fires, to keep the public
roads in repair, and so on. But, should I keep and maintain
these public roads? Was it not better to have them erode and
be abandoned? They were unfit for forestry purposes because
they were poorly constructed and steep, with stony fords in
lieu of bridges and they allowed free access to the moonshiners, the cattle drivers, the hunters, and the prospective
squatters.

Naturally, I was most anxious to purchase the interior holdings held by the bad element of the mountaineers, but these
men in their own interests were most anxious to remain, because moonshining was possible only in the proximity of
cold and remote mountain springs. It was in vain that I
offered a premium for the discovery of any moonshine still.
If one was discovered accidentally and the owner knew it, the
still was removed before I could get the sheriff to seize it.
And if the sheriff actually seized a still and took it home with
him, he never fully destroyed it; he merely shot some holes
through the copper kettle with his revolver and allowed the
kettle to be stolen from his yard in the hope of seizing it a
second time to get a second reward from the federal government.

Both the better element of the mountaineers and the sheriff
were afraid of the moonshiners. I was warned repeatedly that
they would kill me and that I should never go unarmed.
Therefore I carried my Stevens rifle, a wedding present from
Pinchot, dangling in a leather case attached to my saddle. But
I soon abandoned the idea; I was no match for a mountaineer.
His gun was ready sooner than mine, and I was safer, I felt,
going unarmed and making it known that I was unarmed. I
was in danger but twice. The first time was in Brevard, the
seat of Transylvania County, when I returned a horse to its
owner after trying it out for purchase. The owner was much
incensed, claiming that I had lamed the horse, which had

lost a shoe. In his rage he swung his ax over me from behind. I did not dodge, because I did not see it. From that moment on I was considered in Brevard to be a particularly brave man.

The second incident occurred when I saw a man quietly fishing in Davidson River, his horse with buggy hitched to a tree on the side of the public road. When I asked him to stop fishing, he swore at me: "Go to hell!" I replied that I would rather go to Brevard in his buggy and summon the sheriff. He continued fishing. I climbed into his buggy and proceeded slowly on the road to Brevard. After a few minutes, the fisherman caught up with me, his revolver drawn, crying: "Pull out your gun, pull out your gun!" I told him that I did not have a gun; but I was cowed enough to leave the buggy and then at the point of his gun to walk slowly away. I notified the sheriff later, but he, afraid like myself, never went to arrest the man.

With the better class Mrs. Schenck and I soon became friendly. The friendship began when a preacher who was expected at the schoolhouse in the Pinkbeds failed to arrive, perhaps because the river was swollen and unfordable at the time, and I read the congregation a section of the Bible, explaining, as best I could, the text in question and also singing their anthems with them. To improve the singing, I bought a small Windsor organ for $28.75 from Montgomery Ward and Company in Chicago. I played on it, also as best I could, accompanying their singing. Religion, it seems, is a good thing even in a forestry program. And singing is the best peacemaker. I did not know it, but a Methodist preacher one Sunday, having the peace of the world for his text, proved it to me by stating: "There are three great pacifiers: law, religion, and music. Of these three," so he insisted, "music is the most efficient."

My white crews of workmen, in contrast with the black crews, never sang while they worked. The Negroes, however, sang religious songs, apparently because they knew no others. "Jesus, Lover of My Soul" accompanied every stroke of many a Negro's hammer driving steel in road building or every thrust of the shovel in moving dirt.

There were a number of small portable circular-saw mills in Transylvania County, the owners of which I thought I might interest in buying logs from Pisgah Forest, thereby making it possible for me to practice some forestry and also to give employment to the mountaineers. My first experimental log yard was established on Poplar Hill in the midst of the Pinkbeds. Three or four owners of the portable mills were notified and asked to inspect the yard and make their bids in writing. None came, neither millman nor bid. There were the logs on Poplar Hill, and logs once cut do not improve when left in the woods. Thus I was forced to bring up from Biltmore the small portable sawmill which Pinchot had secured and temporarily used on the Biltmore Estate. We hired a sawyer and began to saw, to pile the lumber obtained, and to look for buyers of it. The receipts did not cover the expense. The cost of transportation of the lumber over miserable public roads to the nearest railroad station, now Ecusta, close to the mouth of Davidson River, was far too high. Again — roads, roads, good roads are needed to practice forestry!

At that time Vanderbilt, accompanied by some acquaintances, came to the Pinkbeds on a tenting trip. A trip of that kind was a heavy burden on me. First, I had to erect the tents. They were beautiful, huge tents, some five of them, for the gentlemen, for the Negro camp cook and his helper, for the white muleteers, and for the dining room. Next, I had to secure and transport the necessary supplies, most of which came from Biltmore. Finally, I had to furnish the good weather required to make such a trip enjoyable, a requirement the hardest to fulfill. Among Vanderbilt's guests on this occasion were two bankers from Baltimore, who, after a visit to Lookingglass Rock and Lookingglass Falls, asked me to give them a lecture on forestry as an investment. I was delighted to do so, pointing out that forestry was a factory of trees based on an investment in trees, roads, logging facilities, teams, wagons, houses for the employees, sawmills, lumber-yards, and so on; that, since the trees do not grow at a better rate, the investment yielded a dividend of about three per cent, which could be depended upon annually because the trees would grow as sure as the sun shone and the rain fell. When I had finished my listeners laughed. An investment in America, they said, must yield a dividend of six per cent at least; the very joy of an investment is its mutability, its fluctuation, its risks; and, when an investment declines, the good financier abandons it and applies the funds withdrawn to some other venture.

Vanderbilt was a silent listener. I was nonplused. Little did I foresee at the time that, in a country developing at rapid strides, the dividend in forestry is supplied automatically by the rise in stumpage values much more than by the growth of the trees. But I prophesied to Vanderbilt on another occa-sion that the value of the stand of yellow poplar on the northern slope of Lookingglass Rock, which I had named All Saints Preserve, in time would equal the entire purchase price of Pisgah Forest. My prophecy did not come true. All Saints Preserve was logged outright about the year 1915, when it was sold soon after my dismissal to a lumberman of Italian origin.[1]

Towards the end of 1897 the Southern hardwoods business was in a bad plight, in spite of the gold standard and McKin-ley's election. The business of the Biltmore Lumber Com-pany was losing money heavily. I wanted to wait for a recovery of the market. McNamee, treasurer of the company, wanted to sell lumber in spite of all. Some harsh words and, what is worse, some harsh letters were exchanged between us.

[1] Louis Carr of the Carr Lumber Company.

And, since McNamee was the general manager of the estate and in a sense my superior, I resigned the presidency of the company — a company doomed to failure to begin with. None of its officers knew the lumber business; and no logs were available in Pisgah Forest to feed the sawmill when those of Big Creek had been made into lumber. My resignation was prompted, however, as much by McNamee's arrogance as by our disagreement over the sales policy of the lumber company.

CHAPTER 9

A Trip to Minnesota

ANOTHER INCIDENT OCCURRED at that time that gave me a great shock. I had sold to a group of local persons two hundred acres of outlying lands for five hundred dollars. They were worthless lands, separated from the body of Pisgah Forest by some two miles. A few weeks later I chanced to meet a column of wagons pulling heavy machinery in the direction of those lands. I learned on inquiry that the machinery was for gold mining! Lordy! Gold in Pisgah Forest? Naturally, I was much excited. Had I sold a Vanderbilt gold mine for five hundred dollars? Soon afterward I learned, to my great relief, that the enterprise had failed. The Geological Survey of North Carolina did not encourage any further search for gold. In the geologic formation of gneiss underlying Pisgah Forest, the geologists said, gold nuggets were found accidentally, but rarely. Indeed, no such nuggets have been discovered since then, so far as I know.

I was anxious, of course, to develop the mineral resources of Pisgah Forest if there were any. Permanent roads would be needed for mining, and simultaneously they would serve the purposes of forestry. In the last analysis, the difference between ephemeral lumbering and forestry lies in the ephemeral character of lumber roads and in the permanence of forestry roads. An attempt was made to open a mica mine in Pisgah Forest, but it was not successful. I netted a few hundred dollars for my forestry department; that was all.

In 1898 I chanced to become acquainted with Thomas Alva Edison, then a next-door neighbor of my cousin, George Merck, in Llewellyn Park, West Orange, New Jersey. Edison was deeply interested in radium at the time. Merck and I were shown some of his laboratory experiments. One of his collaborators had been killed by radium, but the victim's brother, inspired by Edison's personality, at once took the place of the deceased chemist. Edison was of the opinion that in the archaic rocks of the southern Appalachians minerals bearing radium could be found in workable quantities. I was very sanguine, since radium from Pisgah Forest would solve my chief problem, that of permanent roads, the construction of which for !umbering alone was not justified at the time. Thus I searched for radium everywhere, sending rock samples to Edison. Unfortunately, only small traces of radium were found in them.

Thinking of Edison, I cannot forego telling of an evening spent as a dinner guest at his house. During the meal he jumped up suddenly and left the dining room for half an hour. "You must excuse him," said Mrs. Edison. "My husband does that continually when he has a new idea; he cannot stand any delay; he must try it out at once in his private laboratory."

After dinner, wishing to pay him a compliment, I assured him that all Europe was admiring him immensely, not only for his great discoveries, but also because they were made by a man without a college education. "Hear, hear!" he exclaimed: "If I had had any college education, I would not have achieved anything, because I would have seen everywhere the limits rather than the possibilities of achievements."

I was deeply impressed; and I must confess that I have

often cursed my own education in forestry, obtained at German universities, when I found that it had misled me, had wrapped me in prejudices or in preconceived ideas unsuited for American application. The Swedes have had some experience similar to mine. Forest working plans made after the Saxony pattern by Swedes who had studied in Tharandt, Saxony, were found to be impossible of execution in Sweden. Of course nobody had learned what "German forestry" Pinchot and I had "planted" in Pisgah Forest, and our good names were left intact. More than that, they had become known in the United States to people interested in forestry.

Thus it happened that the brand-new state forestry board of Minnesota at its first session invited me, through its secretary, General Christopher C. Andrews, to come and discuss its forestry problems on a trip of inspection, with all expenses paid. Members of that board were, I was told, men designated by the Minnesota Horticultural Society, the State Forestry Association, the State Agricultural Society, the State Fish and Game Commission, and the regents of the University of Minnesota, the latter selecting three members, a nurseryman, a farmer, and a lumberman. The lumberman was Frederick Weyerhaeuser. General Andrews, as chief fire warden of the state, and Samuel B. Green, professor of horticulture and forestry at the university, were ex-officio members of the board. Poor me, a youngster absolutely unacquainted with the economics of the United States, consulted, for reasons unknown, instead of Bernhard Fernow or Gifford Pinchot! I accepted cheerfully, with George Vanderbilt's kind permission, and I had a most interesting tour through the cutover lands of northern Minnesota, accompanied by General Andrews and Professor Green.

In St. Paul I was presented to former Governor John S. Pillsbury, with whom I discussed the policy to be applied to the lands which the federal government had ceded to the state for the benefit of its schools. These lands, stocked with forests, were being sold at auction or otherwise, the receipts going mainly to the state university. I argued that the state was neglecting a great opportunity to obtain revenue in perpetuity for the university by disposing of land and timber, instead of selling merely the mature and overmature trees and retaining in the state the title to the lands and managing them for growing continuous crops of timber. Governor Pillsbury listened, but he could not be persuaded. He personally owned large timberlands which he sold in fee simple to the lumbermen at the best price he could obtain. He did not see, or he would not recognize, that there is a difference between investments in timberlands held privately and investments held by public institutions.

At the university Professor Green showed me his nursery. There were seedlings of black locust. He told me, to my surprise, that he was in the habit of pouring boiling water over the locust seeds on the day before they were planted. And there were red cedar seedlings, the seeds of which, he said, had been bruised and left to ferment in wet wood ashes for a number of weeks before being stratified during winter and planted in the ensuing spring. Professor Green knew more than I did, more than my German professors. The seedlings, I was told, were to be used for experimental planting on the Minnesota prairies, of the existence of which I had no previous knowledge.

In Minneapolis I was introduced to a leading lumberman, Thomas B. Walker, who had made a large fortune in white pine and who later founded the Red River Lumber Company in Westwood and San Francisco, California. He entertained us at a luncheon and showed me his picture gallery containing, among other good and bad pictures, two fine Schreyers. All the pictures had been selected, it seemed to me, for the subjects represented rather than for their painters' skill. As to forestry, he evinced no interest whatsoever.

These and other visits over, Andrews, Green, and I went

on a trip of inspection in northern Minnesota, with Grand Rapids for a starting point and Walker for its terminus. What we saw were devastated lands, with miserably poor farms on some of them, and some capital stands of jack pine. I was deeply impressed by the latter. Here was a species which was destined, I thought, to play a role not merely in Minnesota, but also in Germany; a species of pines more modest than any I had ever seen and one which might conquer the very Saharas of the world; a species which was helped rather than ruined by fires and which regenerated itself from self-sown seeds in even-aged stands more easily than any species I had ever seen.

Incidentally, General Andrews visited some of the forest fire wardens of whom he was the chief. These men were local farmers, who were paid for the time they spent in extinguishing, but not in preventing, forest fires. Most of the wardens were also presidents of local village councils or supervisors of towns. They were empowered to summon help when a fire was raging, and any citizen refusing to help was subject to a fine of one hundred dollars. The railroads were required to supply their locomotives with efficient spark arresters and to keep their rights of way, to the width of one hundred feet, clear of combustible material. I was told that this Minnesota forest fire law was the best yet enacted by any state in the Union, and that its efficiency was handicapped only by lack of funds.

On the last day of our trip, while taking a much needed bath in a small creek near our nighting place, against the warning of both Andrews and Green, I caught a severe cold. Enriched by this and other experiences, I returned home and concocted, in accordance with previous arrangement and obligation, a report on forestry for Minnesota. Good Lord! I had seen less than one per cent of the state, and I wrote a report which, no doubt, was utterly worthless for all intents and purposes.

I have before me now a copy of this report.[1] It was addressed to the Minnesota State Forestry Board, of which Judson C. Cross of Minneapolis was president and Greenleaf
Clark of St. Paul was vice-president. The name of Frederick
Weyerhaeuser appears as a member of the board at that time.
My report called boldly for six new state laws and gave the
reasons and explanations meant to prove the necessity for
their enactment. Here is my sextette:

1. A law ordering a survey to be made, by which the nonagricultural townships and sections of townships shall be
defined.

2. A law designed to clear — through the medium of courts,
attorneys and surveyors — the state's title to land forfeited for
nonpayment of taxes.

3. A law creating an organized staff of forest guards, whose
duty it shall be to efficiently protect all private forests, and to
enforce the laws enacted relative to federal and state forests
and forest lands. This staff might, at the same time, have the
duties of fish and game wardens.

4. A law regulating the assessment of nonagricultural land
(compare No. 1) for taxation purposes.

5. A law allowing companies to own tracts comprising over
5,000 acres for forestry purposes.

6. A law providing means for investigating the financial
possibilities of forestry.

On the return trip to Biltmore I spent a day in Chicago.
While there, had I been wise and more courageous, I might
have called on William B. Judson, editor of the *Northwestern
Lumberman*, the leading lumber periodical of the United
States and a stanch adversary of the forestry movement,
which voiced, no doubt, the general attitude of the lumber
fraternity. At that time American lumbermen seemed to
view forestry as an interference with their constitutional
liberties. Why should trees be raised, outside the prairies,
when there were so many trees still at hand everywhere? The

[1] The report is incorporated in the *Fifth Annual Report of the Chief
Fire Warden of Minnesota*, 1899, pp. 125–136.

South was scarcely touched, and the West was totally intact. Had I made the call in question, I might have impressed the editor with the fact that every forester known to me in Europe was also a logging foreman, although none of them was engaged in sawmilling and lumber selling. I might have said that forestry, like any other business, is essentially a problem of transportation, a problem requiring wise national legislation to make it possible for constructive lumbering to be more remunerative than "cut-out-and-get-out" lumbering. And I might have added my conviction, then maturing, that German, Swedish, and other European forestry was not applicable to America any more than were American methods of lumbering applicable to those countries; that America required American forestry, one kind for the prairies, another kind for the East, and a third kind for the West. And I might have admitted that Pinchot and I had made a mess of it in Pisgah Forest!

CHAPTER 10

A Mistress
Arrives at Biltmore

IN APRIL, 1898, I WAS SUDDENLY called back to Germany. My oldest brother had died after a protracted illness. I had no difficulty, under my contract, in obtaining Vanderbilt's permission to take a long vacation. I used my sojourn in Europe to serve one of those military terms to which a German lieutenant of the reserve army was in honor pledged. I used the vacation also to obtain from the German Forest Service, from which I had been furloughed for three years, an extension of my leave for an additional two years.

My long absence from the United States came at a very inopportune time. After two years of lecturing for the benefit of my forestry apprentices, I was ready to carry out my idea of establishing a forest school at Biltmore. With Vanderbilt's consent and Pinchot's encouragement, I had advertised in lumber and forestry papers that a school of forestry was about to be opened at Biltmore, and I had invited applications for admission. My departure left my office staff in Biltmore in a quandary. It could not reply to applicants that I had gone to Germany and that the date of my return was uncertain; nor could it permit applicants to come to Biltmore before I had returned. The result was that the Biltmore Forest School, the first of its kind in the United States, began its activities in the fall of 1898 with fewer students than I had had apprentices in the preceding years. There were daily lectures in my office from 9:30 to 11:30, followed in the afternoons by trips with the students on horseback to the sections of the Biltmore Estate where my work was going on or where work for the coming winter was planned. Taught by my experience with Sir Dietrich Brandis, I had made it a rule — and I stuck to it as long as there was any Biltmore Forest School — that the forestry students should be instructed mainly in localities where things were being done.

When the school was founded, I voluntarily promised Vanderbilt that, since my time was paid for by him, I would turn over to him half of any surplus receipts that I might obtain from the school or from any other work done by me outside the estate. The latter work contemplated expert advice in forestry, expert services at courts of law, or services as an agent of forestry with the federal Division of Forestry, of which Pinchot had just become the new chief, succeeding Fernow.

While I sympathized with Fernow and with his disappointment at the time, I was soon forced to recognize that the change had been for the best interests of United States forestry and, incidentally for my own best interests, for Pinchot made me an agent in his division and consulted me frequently, something Fernow had not done. Fernow, with all his superior eloquence and his glittering personality, had not got far during his twelve years as chief of the Division of Forestry. Pinchot, however, was known to be financially independent; he was no federal servant when he had to appear before a Congressional committee in the interests of forestry; he was a peer of any member of any committee; he could receive in his home in Washington any senator as a guest if he but wanted to invite him; and he was personally acquainted with many senators and other political leaders. Instead of writing voluminous reports on one phase or another of for-

estry, Pinchot began to report on some practical work done
by his division. No wonder that the appropriations of the
Division of Forestry rose under him within three years from
about twenty thousand to almost two hundred thousand
dollars.

The event most important in 1898 for the Biltmore Estate
and, incidentally for Mrs. Schenck and me, however, was not
the official founding of the Biltmore Forest School. It was the
arrival in Biltmore on October 1, 1898, of the new and young
Mrs. George Vanderbilt. Hurrah! Biltmore House had obtained a mistress, and the Biltmore Estate had got a comple-
ment of those qualities which George Vanderbilt manifestly
lacked. We were told that the bride's maiden name was Edith
Stuyvesant Dresser and that the Stuyvesants were one of the
old aristocratic families in New York descending from Governor Peter Stuyvesant; that Edith and her sisters, left or-
phans and penniless, had been reared in Paris under the tutelage of an aunt, good Mrs. Edward King; and that it was in
Paris that Vanderbilt had met his bride.

The young couple, our lord and his lady, arrived on a
gorgeous afternoon and were welcomed by the officials of the
estate at a huge horseshoe of flowers overarching the approach
road to Biltmore House. All was cheers and smiles and hap-
piness! Mrs. Vanderbilt! One could not help but love her!
Her face, by no means regular and beautiful, was sparkling
with kindness, sweetness, lovability, grace, and womanliness.
And I might interject here that Mrs. Vanderbilt's presence
made Mrs. Schenck's life doubly enjoyable, interesting, and
comfortable. Not that she had any direct influence on the
affairs of the estate. The household at Biltmore House continued to be directed by her husband, but Mrs. Vanderbilt
took a personal interest in every man and woman connected
with the estate. She had that fine social instinct that her husband unfortunately lacked; she went to every humble cabin
of a forest worker in Pisgah Forest when she was near it; she

encouraged the small home industries; she shook hands, dirty though the hands might be, with every ranger and ranger's wife; at Christmas she had a small gift for every child of every employee of the estate; and whatever she did or said, she did or said so gracefully that she put at ease whomever she met.

Mrs. Schenck and I were welcomed at Biltmore House and often were guests at lunch or dinner. In addition, the Vanderbilts were with us on all their trips and stays in Pisgah Forest, and the result was a mutual understanding akin to friendship. On one of her first trips to Pisgah Forest, Mrs. Vanderbilt attended a divine service in the Pinkbeds led by me — I do not remember my text. When I had concluded, she exclaimed: "Dr. Schenck, you are a wonder," and she gave me for a souvenir a fine copy of an English Bible, a copy which I read to the present day with very special pleasure.

What of my new forest school? Some of the students were almost as old as I. There was Alfred Gaskill, later state forester of New Jersey; there was Dr. Herbert K. Porter, a real M.D. who abandoned medicine for forestry; there were Albert H. Pierson and Gordon T. Backus, both of whom later served for many years in the United States Forest Service. Overton Price at that time was in Germany studying forestry under Sir Dietrich Brandis and preparing for the position of secretary to Gifford Pinchot immediately after his return from Europe. I was much elated when Pinchot wrote me in 1899 that my graduates had passed the new federal examinations in forestry, some of them at once obtaining positions in his Division of Forestry. My own advice to my graduates, however, invariably was: "Get some practical experience in the woods, if possible with a lumber firm, before you enter the official service of the federal government." What theories they had learned in my school, within a single year's time, were not foundation enough for satisfactory work in forestry throughout the United States.

In the fall of 1898 I was invited by the University of the

Chestnut Wood Chute in Pisgah Forest

Lookingglass Creek

Brandon Hill Plantation
Planted by Schenck in 1905 and thinned in 1948.

South in Sewanee, Tennessee, to inspect its woodlands. Some universities, notably Yale and Cornell, were planning at the time to offer forestry courses, and Sewanee, small though it was, was ambitious to outrival them, believing that it was particularly qualified for forestry by its control of some ten thousand acres of surrounding forests. They were indeed glorious virgin forests, on a nonagricultural formation of limestone rock, well watered by noisy subterranean brooks. I made an address on forestry in the university auditorium and promised to submit to the authorities in the course of the next six months, free of charge, a working plan for the Sewanee University forests. The working plan was made, duly submitted, and discussed, but it was never executed, since the university was short of money. Forestry is no go with an owner short of money.

By that time I had also finished my inventory of all trees in Pisgah Forest having a diameter of eighteen inches or more. The inventory was taken by three good woodsmen assisted by my rangers. It included all sound and unsound trees, and it was meant to give for each tree its species, its diameter breast high, the number of logs in it, and its contents in board feet. The cruisers, axes in hand, tapped the tree with the ax for soundness, and held the ax horizontally against the tree and called their findings aloud to the recorder, three cruisers and one recorder working together as a crew. Soon it dawned on me that the cruiser might read directly from the ax handle the exact diameter of a tree in inches if figures marked on the handles gave, instead of the number of inches covered by the tree seen on a tangent, the diameter of the tree belonging to that tangent. The "Biltmore stick," since used widely in parts of the United States, was thus invented.

A copy of my working plan for the management and development of Pisgah Forest, covering seventy-three foolscap pages, lies before me as I write. It was submitted to George

Vanderbilt on October 1, 1898. It is, I believe, the first real forest working plan ever made in the United States, and, if that is so, it may be of sufficient historical interest to warrant a brief summation:

I. *General Considerations.* The capital invested in Pisgah Forest is $224,880, including the purchase price, all fees and expenses, all taxes, and all outlay for surveys, demarcation, roads, trails, houses, administration, etc., incurred since the tract was purchased. The object of the working plan is to find that stage of capital investment at which the property will yield the highest annual rate of interest. After European experiences, a revenue better than three per cent cannot be expected from a mere investment in trees.

II. *Present Resources of Pisgah Forest.* These resources consist of timber, farms, pastures, and minerals. The timber is very defective: defective are ninety-five per cent of the chestnuts, eighty-five per cent of the poplars, fifty per cent of the white oaks, forty per cent of the red oaks, and twenty per cent of the hemlocks. There is no walnut nor cherry on the tract.

III. *Future Treatment of Pisgah Forest.* Every acre must be devoted to that use and production which will yield the best financial results. The possibilities are farming (there is room for fifty families, the men working during winter on logging); stone quarries and mines; fenced stock pastures under a lease system on thirty thousand acres unfit for tree growth; game and fish licenses for sportsmen; and, of course, growing and cutting timber. Mature trees are those growing at a financial rate of less than three per cent:

SPECIES	NUMBER	CONTENTS IN BOARD FEET, DOYLE SCALE
Poplars	16,845	7,809,900
Chestnuts	28,128	7,795,600
Hemlocks	2,597	1,798,500
Oaks	12,764	4,769,850
Other species	9,301	2,205,000

The aggregate value of these trees is estimated to be $36,560, a fraction only of the purchase price paid for Pisgah Forest.

Immature trees are those growing at a rate better than three per cent, financially, even though they have a diameter exceeding eighteen inches:

SPECIES	NUMBER	VALUE
Poplars	38,791	$ 6,982
Chestnuts	31,224	3,746
Hemlocks	9,562	572
Oaks	55,513	13,322
Other species	16,000	1,920

The annual yield of their growth is estimated to be worth $800.

It is impossible to give a figure for the value of the annual growth of the seedlings, saplings, poles, and trees less than eighteen inches in diameter breast high. During the last hundred years fires have prevented the development of new generations of sound trees. Henceforth the development of such trees must be encouraged by every means, the expense to be charged against capital invested.

Lumber prices are likely to increase at a rate keeping pace with the exhaustion of the virgin forests. At the same time, freight will be cheapened by increased competition and improved methods of transportation. As a consequence, the age and the size of a tree considered to be mature will be reduced. Thus, while a poplar is considered mature when it is twenty-seven inches in diameter breast high, after another twenty years a poplar of, say, twenty inches will reach maturity. Therefore, within a score or two of years, much heavier cuttings will be permissible financially, but only if sufficient regeneration has been secured.

Additional revenue will be obtainable from leasing pastures, hunting rights, quarries, and summer and health resort cottages. The cottages are to be built and humbly furnished at $450 each and to be leased during summer at $25 a month.

The guests will use the products of the farms, fuel wood, and inferior grades of lumber which cannot stand transportation to the railroad.

IV. *The System of Roads in Pisgah Forest.* Alternates to roads are driving and floating, and forest railways, such as cable roads, etc. Wagon roads are recommended because they serve all purposes, but they should not be constructed to begin with in valleys riddled by interior holdings. Such holdings are to be acquired beforehand. The cost of the road system desired would be about $35,900, which, if spread over twenty years, would call for annual outlays of about $1,795. The cost of purchasing the interior holdings is estimated at $20,000.

V. *Budget and Investment Status during the Years 1899 to 1919.* The gross revenue to be obtained in Pisgah Forest is estimated in the following table:

SOURCE	ANNUAL REVENUE	REVENUE DURING TWENTY-YEAR PERIOD
Sales of mature timber (at $1.50 stumpage)	$1,827.93	$ 36,560
Sales of land	771.50	15,430
Pasture leases (during 10 years)	1,000.00	10,000
Farm leases (during 10 years)	500.00	5,000
Sportsmen's leases	2,250.00	45,000
Summer colonies	800.00	12,000
Mineral leases	200.00	4,000
TOTALS	$7,349.43	$127,990

The gross expenses (investments) foreseen and to be made during the years 1899–1919 are as follows:

ACTUAL ADDITIONAL INVESTMENTS	
Roads	$ 35,900
Stock fences	9,000
Boundary fences	1,000

Purchase of interior holdings	18,690
8 cabins for tenants	2,400
10 summer cottages	4,500
Lawyers' fees and resurveys	7,000
Deer to be turned loose	1,250
Cleaning and weeding for regeneration in advance	5,000
Securing workmen families	3,000
Additional trails	2,000

TAXES

Transylvania County, 55,946 acres	18,000
Henderson County, 17,266 acres	3,600
Haywood County, 1,000 acres	2,200
Buncombe County, 7,513 acres	1,800

SALARIES, ADMINISTRATIVE AND
PROTECTIVE STAFF   60,000

TOTALS $175,340

VI. *Stage of Capital Invested in Pisgah Forest by 1919:*

Capital invested previous to January 1, 1899	$224,880
Compound interest at 3% on this amount for 20 years	181,250
Difference between gross expenses ($175,340) and gross receipts ($127,990) during 20 years	47,350
Compound interest at 3% on this amout for 10 years	16,290

TOTAL INVESTMENT IN 1919 $469,770

So much for my working plan for Pisgah Forest submitted in October, 1898. George Vanderbilt approved it verbally, but not in writing. It was put into effect during the years 1898 to 1909, inclusive. Additional interior holdings were acquired; roads, trails, fences, and cabins were built. And then my rupture with Vanderbilt in 1909 brought it to an unhappy termination. Of this I shall tell in a later chapter.

CHAPTER 11

Random Journeys

IN THE CAPACITY OF AN AGENT of the Division of Forestry, I had opportunities during 1899 to make several forestry trips for Gifford Pinchot. One of these trips was to Walker County, Alabama, where I went with all my students. There, in a fine tract of primeval longleaf pine, which was being lumbered, we were to make a study of the rate of growth of that species by counting the rings of the cut trees. The information was to be used by Pinchot in preparing a working plan for the tract.

I do not know what became of the mass of data we gathered. The results were never published. I suppose they were disappointing. The rate of growth of the longleaf pine is so slow that the periods of waiting between consecutive cuts are long. And what was the use to tell a lumberman in a Pinchot working plan that, after a first cut of, say five thousand feet an acre, he would have to wait about a hundred and fifty years before another cut could be expected? I believe our job was not very satisfactorily done. But I myself was highly pleased with what I had learned in logging, in orcharding for naval stores, and in the silviculture of longleaf pine. The students, too, had learned much in those two weeks of no lectures.

In the summer of 1899 Pinchot sent me on a trip of exploration, during which I touched Sumter, Columbia, and Charleston, South Carolina, on my way south. I saw for the first time the coastal plains, the swamps, and some of the Southern harbors, including that of Mobile, Alabama, where I called on Charles Mohr, who had written for Fernow a good report on three species of Southern pines. He was a pharmacist by trade and a botanist by predilection. When I met him he was an old and broken man, who spoke English with a Württembergian accent, although he had lived in the United States for forty years. He showed me a huge trunk filled with Confederate paper money, tears coming to his eyes when he exhibited that relic of olden times. Between Mobile and Pensacola, Florida, I saw the finest regeneration of longleaf pine that I had ever encountered, and I was nonplused by finding that it had developed on land badly and continually burned over. There were also swamps of bald cypress. I shall never forget a trip on a gorgeous Sunday in a rough canoe, steered by a jet-black Negro through the silent and holy cypress swamps, the trees light green and hung with gray veils of Florida moss.

In September I went to Columbus, Ohio, to deliver an address at the state university and to consult with the professor of botany, who was about to offer some courses in forestry.[1] From there I went to Nehasane Park, the Seward Webb tract in the Adirondacks where Henry S. Graves, Pinchot's best man, showed me his and Pinchot's forestry work. It was difficult for me to appraise it correctly, since I had not seen the property in its primeval state before it was lumbered. All merchantable white pine and spruce had been removed; there was no reproduction other than decrepit spruces some twenty to fifty years old; there were no roads of a permanent character, no fire lanes, no rangers, no watchmen. Pinchot and Graves had forecast that a second cut of

[1] Schenck's address was entitled "The Capitalist and Economic Forestry."

some five thousand feet an acre would be obtainable after another twenty years.

I cannot say that I was favorably impressed; but when, in the evening, I met the family of the owner, Dr. Seward Webb, brother-in-law of George Vanderbilt, I did not hesitate to approve of all that Pinchot and Graves had done. I could not do otherwise in loyalty to Pinchot, who so far had shown me nothing but kindness and appreciation. Much of the Nehasane tract was destroyed some two years later by severe forest fires which fed on the lumbering debris. This was a severe misfortune, not only for the Webbs, but also for Pinchot and Graves, who had failed to make adequate provisions for protection of the area against forest fires.

From Nehasane I proceeded to Chicago, on a queer sort of an invitation. One of the great railroads of the United States, the Great Northern, had shown an active interest in forestry, to my surprise, and had invited all members of the American Congress in Washington, regardless of party affiliations, to participate in a forestry excursion by special trains to northern Minnesota. Some leading journalists also were invited. What was it that prompted the railroad magnates to take an interest in forestry? Was it patriotism? Was it a desire to obtain in perpetuity such freights as might be procured by the woods treated conservatively instead of destructively? No! The novel attitude of the Great Northern was due to some Chicagoans, headed by a well-known lawyer, who planned to have the Chippewa Indian reservations in northern Minnesota, comprising some 830,000 acres of land and water, made into a sportsman's paradise. The lines of the Great Northern Railroad traversed that section of Minnesota; and the railroad was not getting an adequate revenue, either from timber or from human freight, out of those great reservations. Thus the sportsmen as well as the railroad were eager to have the reservations abolished and converted into federal forest preserves.

To that end the old treaties made with the Indians would have to be rescinded by the Congress and the Indians, and new treaties would have to be made. The interests of the Indians, as wards of the nation, were entrusted to the commissioner of Indian affairs in the Department of the Interior, and that dignitary, too, had been invited by the railroad to be its guest on the trip. Our party, I believe, was much less numerous than had been expected. It consisted of about sixty senators and congressmen led by Joseph G. Cannon, who later became speaker of the House. Cannon was accompanied by his daughter, Helen. There were two well-equipped trains for us and everything possible was done by the Great Northern for the enjoyment of its guests.

The guests strolled about in small groups in the woods and in the Indian settlements. I chanced to be walking most of the time with a keen young man who seemed to be deeply interested in forestry and who showed a surprising amount of intelligence. When I asked him the name of the university at which he had studied, he replied, "The University of Hard Knocks." And when I asked in what state that university was situated, he and the bystanders shrieked with laughter. Little did I know that my new friend was Elbert Hubbard, editor of the *Philistine* and chief of the Roycrofters, an American celebrated for his wit, his pen, and his handiwork. A number of letters which I now cherish highly were exchanged between us in subsequent years; and Hubbard paid me a fine compliment when he described the Congressional expedition in the next issue of his *Philistine*.

Our party, in a body, visited a charming little lake on an island in Cass Lake. The little lake was solemnly baptized "Lake Helen" in honor of "Joe" Cannon's daughter. The island was stocked with virgin red pine, and at the feet of the red pine was a good stand of young white pine — one of the most interesting cases of a change of species I have ever seen.

The evenings were spent, when there was a chance, in local

assembly rooms; speeches of all sorts were made, some witty, some on forestry, some on the Indians. The most interesting meeting of the group by far was the last session. We were all seated around a huge fire in the open at night, and we smoked the peace pipe with some Indian chiefs. One of the congressmen, acting as speaker, explained to the chiefs what was wanted of them. They should surrender the greater part of their reservations, and in return their weekly pensions would be increased. The replies of the Indian chiefs were made without any gesticulations or manifestations of interest, and none expressed his feelings in that bloomy language credited to the Indians in the campfire stories of the old English writers. At the end of the session the oldest Indian chief, his broad-brimmed slouch hat in hand, passed from one paleface to another begging alms. I was sorely disappointed. Modernism had killed in that much-honored chief all pride, if he had ever had any.

My hands were more than full when I returned to Biltmore. There were the lumber and the firewood trades to be looked after; there were the roads in course of construction, the lawsuits pending, the land options to be taken up; and there were the lectures to be delivered, two hours a day. From year to year the lectures became easier for me. I had my original lecture notes mimeographed, and I substantiated and improved them year by year. At that time there were no primers on American forestry which my students might have used. Thus I was compelled to hand them my own mimeographed notes for study.

During the one-year course at Biltmore the student was expected to become familiar with all parts and parcels of American forestry. There were no vacations. The actual number of schoolroom lectures at Biltmore might have been favorably compared with that of any university offering forestry courses at that time. Earliest among these was the New York State College of Forestry established in 1898 at Cornell

University and headed by Dr. Fernow. I do not know what Fernow's system was in presenting the various topics, or whether he, too, lectured day by day for two full hours, spending his afternoons with his students in the woods. Whatever his system, some of the men trained by him have done remarkably well in later life.

Naturally none of the sciences connected with forestry, such as physics, mechanics, chemistry, zoology, and mineralogy, could be dwelt upon at Biltmore; nor were there any lectures offered on sociology, or on the principles of law. Thus the Biltmore Forest School was more for the benefit of graduates of other colleges who had had those courses than for youngsters coming to Biltmore directly from high school.

Principles of law! My experience in Biltmore practicing forestry proved to me at that time that a forester ignorant of law is badly handicapped. I often blessed the fact that my health had forced me to study law in Germany and to pass the university law examinations before I took up forestry at Biltmore, where I was overwhelmed with legal problems. There were lawsuits all the time, consultations with attorneys, rectifications of boundary lines, options and deeds for new purchases of interior holdings, and contracts with employees, lumber firms, wagoners, and loggers, all of which required a certain knowledge, not of the laws themselves, but of the limits to which I could proceed without consulting an attorney. The law knowledge acquired in Germany was actually more valuable to me in those days than was my knowledge of the theories of German forestry.

I obtained much help for my Biltmore school from officials of the federal government in Washington and from members of the faculties of the University of North Carolina and other near-by schools. Many of these men were willing to spend their summer vacations with me in the Pinkbeds, helping me in my courses on forestry by lecturing on such subjects as economics, geology, and entomology. Thus there came

Frederick H. Newell, head of the United States Reclamation Service, with lectures on irrigation; Professor Collier Cobb of the University of North Carolina, with lectures on geology; and Professor St. George L. Sioussat of the University of the South, with lectures on political economy. The assistant in the farm department of the Biltmore Estate, Malcolm Ross, gave lectures on farming and stock raising, and on many an afternoon he took the students to observe the actual farm work on the estate.

Those early years of the Biltmore Forest School were extremely busy ones. A practicing forester in those days had to spend hours and hours, not in the woods superintending jobs, but on the roads to the various woods jobs. It was before the day of the automobile, and I had to ride horseback or drive a buggy, frequently combining driving and riding; and I had to walk and climb a great deal, too. Three lusty cheers for the goose step of the German army! It had given me strong and untiring legs!

Gifford Pinchot at that time was engaged in establishing a school of forestry at his beloved Yale University. He showed great wisdom and great unselfishness when he selected for the Yale chair of forestry his right-hand man, Henry Solon Graves, who was also a Yale man and who had spent two years in Germany studying forestry in Munich under Heinrich Mayr and under the eyes of Dietrich Brandis. Mayr told me when I visited him, as I always did on my trips to Europe, that Graves had been the most industrious and hard-working student he had ever known, and that he had passed the Munich examinations with flying colors. Moreover, Graves could look back on his experiences at Biltmore, Nehasane, and elsewhere when he began to give instruction in forestry. And, being American born and a Yale graduate in addition, he was better qualified for a leading role in American scientific forestry than either Fernow or I.

The Yale students, however, suffered under one great disadvantage: there was no forestry going on in the vicinity of Yale. Pinchot's parental estate at Milford, Pennsylvania, was too far from New Haven to be a sufficient substitute for a Yale laboratory of forestry. The Cornell forestry school, on the other hand, obtained a capital working field when, at the suggestion of Governor Frank S. Black, the New York legislature gave Fernow permission to select thirty thousand acres in the Adirondacks for demonstrations of practical forestry. Fernow took a solid tract near Tupper Lake, which was fairly well stocked with trees and contained few burns. Unfortunately, the tract was not easily accessible from Ithaca, and Fernow as well as his students lost much time on their way to and from their forest laboratory. For other reasons, however, Fernow's forestry in the Adirondacks — much like Pinchot's at Nehasane — was to end in a terrible debacle, as I shall relate later.

Neither Pinchot nor Fernow nor I, when we began our operations and demonstrations, had any knowledge of logging and lumbering. Logging, we thought, was easy; no education was needed for it; any old fool could log. We forgot that the number of successful loggers in the United States was very limited. I am free to confess that I was a bad forester in the United States to start with, because of my ignorance as a logger and lumberman. I made every effort, however, to learn by observation; and George Vanderbilt was kind enough to pay the bill for my education during no less than fourteen years.

By 1900 my house on the Biltmore Estate was completed. It would have been hard for me to find anywhere in Germany a better forester's home. Thus, when the German forestry authorities wrote me in the early spring of 1900 that my leave of absence would not be extended any longer and that I should report for duty in Darmstadt, I was much upset. My life and that of my wife in the United States were most interesting, and I was fascinated with the various tasks in my charge. I did not want to return to Germany, but neither did I want to abandon my German fatherland and my German citizenship. Furthermore, in case of my death, Mrs. Schenck would draw a pension if I remained a German forester, and if I met with any disabling accident, I would have a continued and secure income from the German government. And it happened that I was offered, at the same time, a professorship in forestry at the German University of Giessen, where I had obtained my Ph.D., so that my love for the teacher's lot might have found its final goal.

With these points in view, I went to Vanderbilt and told him of my perplexity. Vanderbilt, too, it seemed to me, was upset when I told him that I was tempted to leave and to leave for good. He proposed to give me all the financial inducements that I would lose if I refused to heed the summons of the German forestry officials. A new contract of employment was drawn up, granting me an increase in salary, year by year, up to five thousand dollars, a life insurance annuity of eleven hundred dollars if I should be disabled or if my job at Biltmore should come to an end, and the sum of ten thousand dollars to my wife, should I die while in Vanderbilt's employment.

With this contract in my pocket, I sailed in early April, 1900, this time via Italy, for Europe, where I had to serve another term of two months in the German horse artillery. Some of my students came with me to study German forestry, although I knew that I could not give them much of my time in Germany. For a German-employed forester the army service — a lieutenant's army service — was at that time a glorious vacation and freedom from forestry routine. For me, it was a severe gentleman's duty which I had pledged myself to perform at a time when I had no thought of the United States.

And on this occasion I had bad luck. The captain commanding my battery of four field guns was a young aristocrat who detested anything smelling of democracy or of the United States. He seemed to believe that it was his duty to de-Americanize me. On the first morning of my term, when I was riding in front of the guns over the practice grounds, the captain exclaimed so loudly that every soldier could hear him: "Lieutenant Schenck, honestly, your American style of riding makes me sick. May I ask you to ride behind my guns?"

Now, that reproach was more of an insolence than I could swallow. I did not ride to the rear of the cavalcade. I rode back to the barracks. The captain, returning some hours later, accosted me: "You seem to have misunderstood me when you rode home." "No," I answered, "I rode as far back of your guns as I could, under your own orders."

From that day on my life in the army was made miserable. I did not have a free hour. All routine service was thrown on me, all jobs which any subaltern sergeant might have done better than an officer. When there were drawers or socks to be inspected, Lieutenant Schenck had to be present; when oats for the horses had to be fetched and weighed, Lieutenant Schenck was commissioned for the job; when the sentries had to be controlled at four A.M., Lieutenant Schenck was ordered to do the work; and on Sundays, Lieutenant Schenck had to accompany the men to church. I was in a terrible fix and in a rage still more terrible because I could find no remedy. An officer lodging a complaint against his superior in Germany was always found to be in the wrong.

Worse off, however, were my students, to whom I could give no time. In fact, this visit of 1900 was the saddest visit to the fatherland I ever had, all because of one single overbearing officer who chanced at the time to be my chief. I took the consequences, however, and found means later to withdraw from all obligatory terms in the army from that time on.

I might say here that I never had the impression when I wore a uniform that anybody in the German army was preparing or was being prepared for a war. No, it was all for the show, for parades, for bluffs, for an aristocratic play. The common soldiers went through a training in gymnastics which, since there was no organized public sport in Germany, took the place of football or baseball. And the army was democratic insofar as everybody had to serve in it, the prince alongside the beggar.

My term lasted two full months — as long as my vacation — with the result that I did not see anything in forestry excepting my good friends Heinrich Mayr in Munich and William Schlich, who was busy with his students on a forestry tour through Germany. Of course, I reported for duty at forestry headquarters and begged to have my leave of absence extended for another two years. I had to depart with Mrs. Schenck before the definite answer was received.

Narrow Gauge Railroad in Pisgah Forest

Wicker Fences to Control Erosion on Biltmore Fields
[Photo by U.S. Forest Service]

Three Day Camp, Pisgah Forest

Buckspring Lodge, Mount Pisgah

CHAPTER 12

Some Notables
Who Came to Biltmore

RETURNING WITH MRS. SCHENCK and three German servants on the steamer "Pennsylvania" of the Hamburg-American line from Paris and Boulogne, we celebrated the Fourth of July in mid-ocean. On our arrival in Biltmore on July 7, 1900, I was welcomed by some sad news. Gigantic fires had celebrated my absence from Pisgah Forest; five of the houses in the woods had burned down; eight miles of barbed-wire fences marking our boundary lines, and incidentally keeping cattle out, had been cut to pieces. The afforestations on the abandoned fields, however, looked very good. I had spent on them, as I only then discovered to my great shame, $1,250 more than my budgetary allowance.

And more sad news! There were few new students in view for my Biltmore Forest School. The competition of Cornell, Yale, and some other forestry schools then in the making made itself badly felt. Naturally, a graduate of any of these universities would entertain no preferences for Biltmore, even if his father was a lumberman; and lumbermen at that time were beginning to see that forestry as preached at Biltmore had certain undeniable merits. At Biltmore any activity by the owner in his woods was considered to be forestry; the critic, if it did not please him, might call it bad forestry, but forestry it was. If the assets of all kinds found in the woods were being conserved at their former financial level, Biltmore spoke of it as "conservative forestry"; if they were destroyed, Biltmore called the reductive process "destructive forestry"; and if the budgetary assets were increased, by afforestation or by the construction of permanent roads or fire obstructions, Biltmore spoke of it as "constructive forestry." Also, it was recognized that the owner, under conditions which he could not change singlehanded in a democracy, was frequently forced to elect the course of destructive forestry.

At Biltmore House, on my return from abroad, there were several unusual guests. Among them was a famous obstetrician who took long walks with Mrs. Vanderbilt, and several women said to be midwives. These personages, together with Mrs. Vanderbilt's altered outlines, forecast a great event. An heir for Biltmore was expected! What an event for an estate on which the general manager, the chief clerk, the landscapist, the farm heads, and notably Mrs. Schenck and I, were without children and without any prospects of them! The foremen and the rangers had plenty of children, but the chiefs of the departments had none. And here was a Vanderbilt heir in the making. When the heir — nay, an heiress — appeared on August 22, 1900, she was named Cornelia after her great grandfather. George Vanderbilt, happiest of fathers, was all smiles. There was nothing lacking from the perfection of all his dreams and schemes.

Among the celebrities, not already mentioned, whom I chanced to meet at dinners or at luncheons at Biltmore House, I might mention, in the first place, John S. Sargent, the most famous painter of the time, who painted pictures of old Mr. Olmsted, of a number of Vanderbilt connections, and of George Vanderbilt himself. All were striking likenesses and all were painted with astonishing rapidity, in two or three sittings. Of course, Richard Morris Hunt, the architect of Biltmore House, was there often in discharging his responsibilities for the outer and inner arrangements of the house. Hunt died in 1895, unfortunately, when the house was just completed. His son Richard and I became such close friends on a trip to Buckspring Lodge that we exchanged our habiliments, young Hunt getting my colorful costume of a German forest officer and I receiving his brand-new Paris-tailored waistcoat, double breasted, sky blue, and adding to the wearer's good cheer on festive occasions.

The most amazing acquaintance that I made at Biltmore House was the Chinese ambassador, Wu Ting Fan — amazing by reason of his sparkling witticisms, his equally sparkling silk robes, and his unbelievable tales. The first telephone was made and used, so he said, not in America but in China, in the year 2666 before Christ. A talking box repeating what had been spoken into it was used by Emperor Wu Wang in 1122 B.C. China, he insisted, was on top of the world. But the attitudes and the preconceived ideas of the Chinese and the Americans were so different that they never would understand one another.

This ambassador had refused, so he related, the president's invitation to attend a big horse race in Washington on the grounds that he, the ambassador, already knew that one horse was running faster than the other. He was not in the least astonished when I related that a colleague of his, a former ambassador of China in Paris whom I had met in Menton in 1895, had shown me a huge pile of letters addressed to him and left unopened, because, he said, "they might contain something disagreeable." Nor was he astounded that among the Parisian ambassador's hand luggage was his coffin ready for use.

The most frequent guest at Biltmore House was Paul Leicester Ford, writer of history and fiction. Ford was a dwarf, some four feet tall — the result of an early injury to his spine — but symmetrically built. Looking at Sargent's picture of Vanderbilt's beautiful cousin, Mrs. Edward Bacon, he could not refrain from throwing kisses to her. He was happily married to an equally beautiful girl of normal height. I have today in my library his novels *Janice Meredith* and *Tattle Tales of Cupid,* copies of which he autographed for Mrs. Schenck. Had it not been for his dwarf size, of which he alone seemed to be unconscious, I would rank him among the most bewitching gentleman whom I have ever met, with his flashing wit, keen observation, and world-wide education. He was killed by his own brother, whom their father had disinherited.

George Vanderbilt's brothers and Mrs. Webb, his sister, came to Biltmore very rarely. I believe that, possibly for financial reasons, they were not on the best terms with George, who had been the favorite son of his mother and, I was informed, her main heir. I was never invited to Biltmore House when any visitor with the Vanderbilt name was present, but I was with the Bacons, the Sloans, the Scheffelins, and others of Vanderbilt blood at Biltmore House as well as at Buckspring Lodge. Mrs. Vanderbilt's elder sister Susan, wife of the Viscount Romain d'Osmoy, who had a small country estate in Normandy, lacked the charm of the Lady of Biltmore. The count was a good sportsman with gun or rod and was somewhat unhappy when he had no luck on a deer hunt in the Pinkbeds.

Mrs. Alice Longworth, wife of a wealthy congressman and daughter of Theodore Roosevelt, was a visitor some few years later when, in her honor, an exhibition of carpentries and weavings made in Mrs. Vanderbilt's Biltmore Shops was opened.[1] I cannot say that we made any impression one upon the other.

Much more interesting to me were the Bryces — the English ambassador, Viscount James Bryce and Lady Bryce, his wife. The ambassador at that time was without a doubt the

[1] The shops had been established to aid the mountaineers by fostering native crafts. They are still in existence.

best-known foreigner in the United States. His renown was well merited. It was he who in 1888 had written *The American Commonwealth,* a book followed later on by an equally great one, *Modern Democracies.* Bryce, who was a Scotchman by birth, had been regius professor of civil law at Oxford for twenty-three years, a Whig member of Parliament, and a cabinet member in the liberal ministry of 1892. When the Tories came into power, they were wise enough to keep him as ambassador to the United States. He was by far, I should think, the best educated and therefore the best qualified of all foreign envoys to the United States.

Viscount Bryce was an expert in the history of all countries. He had traveled everywhere, and he had a wonderful memory for everything he had observed, most certainly so for things he had seen in Germany. He was also a geologist, and in botany, I must confess, I was no match for him. I felt utterly ashamed that he knew the botanical names, of which I was totally ignorant, of many herbs that we found on our excursions over the estate. He was eager, in particular, to plant on his Scotch country place the twenty kinds of oaks found at Biltmore and in Pisgah Forest; and I sent him the acorns of most of them in the autumns following his visits.

Ambassador Bryce! What a grand representative he was of Old England! And how badly was Germany represented by her ambassadors in Washington! I now believe that the United States would have stayed out of both the First and Second World Wars if the German ambassadors in Washington had been Bryces, if they had had his command of the English language, his fine understanding of things American, his forebearance and patience with American qualities of an un-European type. I have met in my life only three German ambassadors to the United States. Of one, Theodore von Holleben, I have spoken previously. Another, Speck von Sternburg, did not care to attend that big Rooseveltian congress of forestry to which he had been invited because a representative of the emperor could not attend a meeting which His Majesty would not have attended himself. The third, Dr. Hans Luther, was sent by Hitler to Washington as Germany's ambassador, although he knew neither the language nor the country nor the inhabitants. Luther, however, was, and he still is, a great admirer of Franklin D. Roosevelt, and he becomes eloquent when he eulogizes Roosevelt's political versatility.

Good Lord! How few of the ambassadors of the world are or were workers in sympathy with the doctrines of the Sermon on the Mount: "Blessed are the peacemakers: for they shall be called the children of God."

CHAPTER 13

Foresters, Beware!

THE YEAR 1901 BROUGHT TO western North Carolina some important economic improvements, among them leatherworks, tanneries, tannic-acid plants, veneer works, and furniture factories. The Southern Railway had acquired control of some feeder lines, including the ramshackle string of rusty rails that ran from Hendersonville to Brevard in the counties in which the main body of Pisgah Forest was situated. Through freight rates to all places in the United States were from then on obtainable at the various railroad stations in the proximity of Pisgah Forest. A modern hotel, the Franklin, was built in Brevard, where rooms with real baths were obtainable, and my frequent overnight stays in Brevard ceased to be disturbed by fleas, bedbugs, and rats. Brevard was the seat of Transylvania County, which included the larger part of Pisgah Forest.

One day the president of the Hendersonville and Brevard Railway, J. F. Hays, accompanied by George L. Adams, came to the Pinkbeds of Pisgah Forest with the news that they were looking for a good location for a tannic-acid plant at Brevard. Hays asked me, point-blank, whether or not I was interested in their scheme. Of course I was deeply interested, provided the plant was built within easy reach for deliveries of chestnut wood and tanbark to be converted into tannic acid. To that end it would be necessary to construct a wagon road a mile or two long from Vanderbilt's property through the bottom lands of the French Broad River separating it from the railroad track. The visitors promised to obtain the rights of way necessary for the road, to secure for me some two acres of land at the railroad track for a lumberyard, and to erect the plant on a site as convenient as I could ever hope to attain — all under the condition that the forest department of the Biltmore Estate would supply the new plant for ten years with three thousand cords annually of chestnut wood and one thousand cords of oak-tanbark at fair prices.

This was, so it seemed to me, a chance I had not dared to dream of, a real God-sent opportunity. The great economic drawback of Pisgah Forest had been its inaccessibility; now, of a sudden, a capital access would be made available at practically no cost. Scarcely three years had elapsed from the day when I submitted to Vanderbilt my working plan for Pisgah Forest, alluding in it to the possibility of selling tanbark and tannic wood from Pisgah Forest in years still far ahead. Already an economic revolution was taking place in western North Carolina, and it was the owner and founder of Biltmore who had brought it about by drawing the attention of all America to the area. Vanderbilt approved my recommendation that we accept Hays's proposition.

There was one thing, unfortunately, that I did not know. My working plan had given a fair tally, vale by vale, of the timber resources of Pisgah Forest; but I did not know its resources in tanbark and chestnut tannic wood. There was not time for an exact cruise of the entire area, and I had no volume tables giving the ratios of tree diameters to cords either of chestnut wood or of oak-tanbark. So I canvassed the valley of Davidson River, which was nearest to the prospective site of the tannic-acid plant, for chestnut wood, running some sample strips sixteen feet wide across the valley and calipering the chestnut trees found within the strips. On an acre I had the bark peeled off the chestnut oaks after calipering the trees. The result of these investigations was so satisfactory that I did not hesitate to close the contract to deliver three thousand cords of wood and one thousand cords of bark annually. The prices were very fair, so it seemed to me, leaving a good profit.

I would have closed the contract even if there had been no profit, because it would secure for me, practically free of charge, access to a railroad and a loading yard, and it would furnish employment for the next ten years to the workers living within Pisgah Forest, regardless of the vicissitudes of the lumber market. After closing the contract, however, I found that I had made two serious mistakes: I had overestimated the amount of tanbark available, because my figures of the ratios between tree diameters and weights of their bark were badly overestimated; and I did not think of the likelihood of increased wages when a much increased demand for labor should come about.

I also neglected a third item. Tanners in those days were "skinners"; they did not hesitate to "skin" all parties with whom they dealt. By my contract, I was to deliver all tanbark well dried and free from mold and defects. My bark deliveries were at the mercy of the tanner. A cord of bark is the equivalent of a ton. In many, many cases my purveyors were docked thirty per cent and more. The scales on which the weights were taken were owned by the tanner, and to prevent cheating I was forced in the end to run my deliveries over scales of my own before they reached the tanner's scales.

There was another drawback in my bark contract. Tanbark of oak (chestnut oak, *Quercus montana*) can be easily peeled only during April and May. These two months in western North Carolina are rainy ones, with whole weeks during which it is impossible to dry any bark in the woods. If the bark does not dry out, it is apt to become moldy and worthless. Thus it came about that my tannery contract gave me more worries than any other labor of my entire life.

Foresters of the world! Beware of contracts with tanneries! And particularly beware of contracts stretching over and binding you for as many as ten years. Had I been wise, I might have profited from the experience of Dr. Fernow, whose fifteen-year contract with the Brooklyn Cooperage Company had led him and his Cornell forestry school to disaster shortly before my tannery contract was consummated.

As already mentioned, Fernow in 1898 had obtained a working field of thirty thousand acres in the Adirondacks for his Cornell school. When he took charge of this large tract he was naturally eager to demonstrate in the woods what he was teaching in his classes and what he had been preaching all over the United States for more than ten years: *practical forestry*. His chances for success were not bad: his tract was situated near centers of demand; it was accessible by rail and by water; it was supplied with some good wagon roads made and maintained by the state; and it adjoined lands that were protected from fires by the New York State Commission of Forests, Fish, and Game.

True, the greater part of the mature timber consisted, not of softwoods, but of hardwoods, especially beech. But there was the Brooklyn Cooperage Company, an offspring of the Havemeyer Sugar Company, buying beechwood at fair prices for the manufacture of sugar barrels and willing to pay an extra price for Fernow's beech if he would guarantee deliveries of ten thousand cords of split beechwood for fifteen years. Naturally Fernow thought this a unique chance to convert his overaged beech into money and into promising thickets of spruce and white pine.

From his timber estimates, made with the help of his students, Fernow was sure that his stands of beech would easily satisfy all future demands of the Brooklynites. And he had a fund of thirty thousand dollars, which had been given to his institution by the New York State legislature, from which all preliminary expenses might be defrayed. After the beeches were cut, the few spruces mixed with them would be preserved and left as seed trees; where regeneration of spruce might fail, there were his nurseries at Axton to supply the seedlings needed for planting; and, finally, there were his students at the camps built for them at Axton to do most of the nursery and planting work free of charge.

For barrel staves only strong and sound wood can be used. The question arose as to what should be done with the small, crooked, and unsound pieces of beechwood. To solve this problem, Fernow succeeded in having established near his tract two other branches of wood consumers, a distillery for producing wood (methyl) alcohol and a brick and lime kiln requiring fuel wood. Indeed, he went so far as to plan the utilization of small branch wood, which was to be bundled up and sold for household fuel. Naturally, he knew that the prices he might obtain for the fuel wood would not cover his outlay. The loss was to be charged to silviculture. It was a part of the cost of a second growth of spruce and white pine produced in place of the old decrepit virgin beeches.

Seed trees of spruce were rather rare. But Fernow had planned to rely less on these mother trees for the propagation of a second stand than on artificial planting. Strange to say, he wanted to raise, not the spruce native to the Adirondacks, but the European spruce, although he did not know and could not know whether or not it would stand the winter snow and cold and the summer heat of the Adirondacks. He figured that the cost of planting could be reduced to six, or even five, dollars an acre as soon as his business was fully established. A few hundred acres of old burns were afforested in the first two years.

According to Fernow's plans and to his contracts, some 2,500,000 feet of beech logs, 10,000 cords of stave bolts, and 2,500 cords of kiln fuel wood were to be cut annually. After the logging began, it soon became apparent that ninety-nine per cent of the beeches were unsound, and that when a tree was cut only a portion, and often only a very small portion, of the bole was fit for barrel staves. Moreover, the cut of 10,000 cords of stave bolts was paralleled by a cut of 30,000 cords of fuel wood in excess of what his market could absorb — a result which no one could have foreseen.

Fernow's working capital of thirty thousand dollars, of course, was soon exhausted, and Cornell University did not care to underwrite him. The Brooklyn Cooperage Company insisted on its contract, its factory being ready long before any wood could be obtained to feed it. Without money Fernow could not buy steel rails, but he was lucky enough to obtain the loan of some old and rusty secondhand rails from a firm in the West merely for payment of the freight to the East, where the company hoped to get a better price for the rails as soon as Fernow should be through with them.

But worse luck pursued him. Close to his operations were the summer residences of wealthy and influential New Yorkers. These men saw in his operations the loss of the forest surroundings in which they had been summering for many years. The beauty of nature, they felt, was being destroyed by forestry. One of the outraged summer residents was a lawyer, who happened to remember that a constitutional amendment adopted by New York State in 1894 prohibited any lumbering on the state's forest preserve lands. Fernow's tract was preserve land leased to Cornell University by the state of New York. This lease could not allow what the state constitution forbade. The result was a state court decision stopping all lumbering for the Brooklyn Cooperage Company or for anyone else.

In addition, the Society for the Protection of the Adirondacks raised such a protest against Fernow and his forest school that the state legislature withdrew the annual appropriations for maintenance of the institution. Consequently there was no more money at Cornell for salaries and other needs, and the university, as the host of the institution, was unwilling to come to the rescue. Unfortunately, Fernow himself lost standing with the New York State College of Agriculture at Cornell, with which his state College of Forestry was co-operating, and also with the Cornell University authorities. There was no need for him to resign. He was suddenly out of a job. While he had made a contract of long duration with the Brooklyn Cooperage Company, he had not made any contract for his own employment for any length of time by the College of Forestry. More than that, he was accused under the criminal code of the state of having violated the constitutional prohibition concerning lumbering on the forest preserve lands.

In this predicament, Fernow asked me to help him with a written professional opinion on the merits of the forestry work done by him in the Adirondacks. Whom else might he have asked? Not Gifford Pinchot, nor Charles Sargent, nor the New York State Commissioner of Forests, Fish, and Game, with all of whom he was at loggerheads. I did not like the task; I felt in advance that I might excuse and defend Fernow's actions, but that I could not approve them from the forester's standpoint.

Spending a few days with Fernow at Axton, I was shown about by him. There was snow on the ground and we had to use snowshoes that looked like tennis rackets, something I had never before seen nor worn. Fernow and I, to start with, raced with each other for exercise, both repeatedly falling over our own feet. In the woods, following his railroad right of way, we saw some remnants of piled beechwood and many piles of burned debris. The spruces that had been left for seed trees had been laid flat by a recent storm. Fernow claimed that the loss of the trees had been due to bad luck, since the storm was of unprecedented ferocity. But he should have known that spruce is a species apt to be blown over when the trees are greatly thinned and suddenly exposed to changed weather influences. I was sure that these mother trees, had they not been felled by the wind, would have died during summer from exposure on a soil deprived of protection against evaporation.

In my official opinion, however, I could say with a good conscience in Fernow's defense that failures in the beginning were to be expected in any novel venture; that a working fund of thirty thousand dollars was utterly insufficient to defray the expenses connected with the establishment of experiments on so large a scale; that lumbering was as much a part of forestry as arm and leg are parts of the human body; and that a state constitution forbidding lumbering outlawed forestry.

Fernow was disappointed when he received my written opinion. Apparently he had expected to be whitewashed, for when later on he autographed for me a copy of his book on *Economics of Forestry,* he wrote on its cover page the words: "In memory of many agreements and disagreements." Scientifically, Fernow's reputation had suffered in the United States. But he had saved and conserved, and he took with him to Canada the best of all rewards — the devoted attachment, nay, the love, of all students who ever had attended his lectures at Cornell.

CHAPTER 14

The Year of Upturn

FORESTRY IN THE UNITED STATES took a most decided turn upward in 1901. It was the year of William McKinley's assassination and of Theodore Roosevelt's ascendancy to the White House. No president before Roosevelt had ever lived in the real West of the United States. No president before him had recognized from actual experience the interdependence of forests and waters. Roosevelt had spent two or three years on a ranch near Medora, North Dakota, where the Northern Pacific Railroad crosses from North Dakota into eastern Montana. He also had a close connection with the South through his mother, Martha Bulloch, who was proud of her Southern ancestry. Thus it was that Roosevelt knew from more than hearsay the urgent need of forest conservation for the entire West and the entire South.

Roosevelt abounded with energy; and in his forestry interests he teamed with the best forester in the United States, Gifford Pinchot. The Roosevelt-Pinchot team, brought about by the sad accident of McKinley's assassination, was of the greatest and happiest consequence for American forestry. Aside from his official connection with the president, Pinchot had the opportunity to meet him socially, especially on the tennis courts of the White House and on horseback at riding parties, with the result that the forester of the United States, unacquainted with the problems of "the wild and woolly" West, had the president of the United States for an adviser.

On July 1, 1901, governmental forestry was raised from the humble position of a division of the Department of Agriculture to the rank of a bureau, headed by Pinchot. The new Bureau of Forestry embraced five divisions: forest management, forest mensuration, forest extension, records, and dendrology — the last under the best possible man in the world, George B. Sudworth. True, there was not one man in the Bureau of Forestry who had an all-around training or experience of his own in forestry, logging, or lumbering. Enthusiasm for the task had to make up for this serious deficiency.

About a year after he became president, Roosevelt made a tour of the South, during which he stopped at Asheville, a thoroughly Democratic town. There, on September 9, 1902, he had the courage to speak to masses of his political opponents from a platform in the public square. The square was packed. Roosevelt spoke on three great civic virtues, namely, honesty, courage, and common sense. He spoke in a manner resembling closely that of the German Emperor Wilhelm II, his mouth set square, so that his white teeth appeared between the angles of the squared mouth drawn backward. And he spoke like the Emperor, in short, abrupt sentences pushed out rather violently and interrupted by several seconds of silence. In the midst of his address, a torpedo was exploded. Everyone thought of another attempted assassination, and some people began to flee. Roosevelt alone was entirely undisturbed. He must have had glorious nerves.

When I learned that Roosevelt was visiting the Vanderbilts at Biltmore House, I made it a point to call there, driven partly by curiosity to see a president of the United States at close hand and partly in the hope of discussing with him the problems of American forestry. When he shook hands and learned that I still retained my German citizenship after six years in the United States, he exclaimed: "Nobody has a right to work here for so long without becoming a citizen of the United States!" His words fell on me like a cold shower bath. For a discussion of forestry there was no chance; the president of the United States did not care to be advised by an alien. Why should I change my citizenship while staying in the United States when no American staying in Germany was abandoning his affiliation? I was proud to be a German.

One of my jobs at this time was forester for Highland Forest in Jackson County, North Carolina, on the waters of Tuckaseegee River, a tributary of the Tennessee River. The tract contained some forty-five thousand acres and adjoined Pisgah Forest along a stretch of some ten miles. Its chief value consisted of minerals rather than trees. On Sugarloaf Mountain there was a mine of ruby corundum, the hardest mineral next to diamond, used exclusively for emery wheels. The boom collapsed when artificial corundum took the place of the mineral, and the bankers who had backed the mining enterprise lost much money. In an attempt to minimize their losses they decided to develop the forest resources of the property. I was offered the job of forester at a salary of a thousand dollars yearly, which I divided with Vanderbilt, as I did all my professional revenues.

Highland Forest had more interior holdings than Pisgah Forest; its boundary lines and its titles were even more in dispute. My chief tasks, consequently, were those of a lawyer and a surveyor. Interior holdings had to be purchased in order to solidify the tract and to make possible the utilization of its timber. And a map had to be made showing the geography, the geology, the timber, and the sites of the minerals, which included rum-colored mica, kaolin, and soapstone, and there were indications of copper and nickel.

This task, while it had nothing to do with forestry proper, fitted into my forestry program. A forester must be capable of developing all resources of the property in his charge. As helpers I used some of the best graduates of the Biltmore Forest School, Gordon T. Backus and, after him, John Lafon. Their work was admirable. Without my graduates I would have been helpless on the Biltmore Estate, in Pisgah Forest, in Highland Forest, at Sewanee, on the holdings of the Tennessee Copper Company, and in the expanding work of C. A. Schenck and Company.[1]

The Tennessee Copper Company in those early days was the largest producer of copper in the United States. Its mines and smelters were situated at Ducktown in Tennessee, close to the point where Tennessee, Georgia, and North Carolina meet. Its owners were Adolph Lewisohn and Sons of New York City. Its attorneys, with whom alone I had to deal, were a firm in Chattanooga, Tennessee. There being no coal, wood of any kind was used in treating the ores. They were treated in open smelters, from which arose a continuous stream of gases, notably sulphurous acid gas, known to be deadly to all plants but not to animals and men. All dead timber had been made into fuel wood for the smelters, and the requirements of wood were so great that all timberlands in the proximity had been exhausted. As a consequence, the entire surroundings had been converted into a desert dissected by deep gullies and ravines, looking like a landscape on the moon.

The copper works were employing many thousands of hands, not only from Tennessee but from the adjoining states of Georgia and North Carolina, in the mines, in the smelters, and in cutting and transporting wood. The wave of death to all vegetation had been spreading southward, deep into Georgia; many Georgians were suing for damages inflicted on their crops and woodlands; and the state of Georgia was suing the state of Tennessee in the United States Supreme Court to stop a public nuisance against which the neighboring state was defenseless.

[1] The company was established by Schenck to act as consultant foresters. Schenck employed some of his former students in the work.

Here again I was consulted in a legal problem connected with forestry. It happened to be a problem of which I chanced to know a little something, because my father, as an attorney at law and simultaneously a member of the directorate in a chemical firm at Mannheim, had had an analogous case to deal with while I was a student of forestry in Germany. There it was the town forest of Mannheim that was alleged to be suffering from the poison smoke emitted by the chimneys of the chemical works. In the Mannheim case, I had found in the dead trees huge colonies of bark beetles. The courts, being unable to decide whether the death of the trees was due to beetles or to poisonous gases, found for the defendant.

Remembering all this, I secured the German literature on the subject and went to work at Ducktown with the help of my graduates. In eight directions of the compass, radially from Ducktown, they collected at intervals of five hundred feet the leaves of the various species of trees and the insects found living on them. The leaves were sent to the chemist of the University of Tennessee for an analysis of their sulphurous acid content. The species of the insects I determined myself, and I soon found that the damage done was in no wise connected with insect work. It was made evident, however, that there was not a foot of woodlands that had not been frightfully ravaged annually by severe forest fires. There was not a tree whose bark was not split open close to the soil and whose trunk was not blackened by fire. Thus the chemical analyses gave me the surplus contents of poison (a small amount of sulphurous acid is normal in all leaves) and an index of the severity of the gas attack on the various species of trees.

From these data I could show that the vitality of the trees had been most severely handicapped by the neglect of the owners to protect their lands against fire and by the failure of the states of Tennessee and Georgia to enact proper forest fire laws. I found, in addition, that the susceptibility of the trees depended on their species, white pine being most sensitive and black gum least sensitive. My final report, made after several years of investigation, claimed that sulphur fumes could not be held responsible for the death of trees within a "smoke region" if (1) the tree species known to be more sensitive is suffering less than the species known to be more resistant; (2) if tall trees are less affected than small ones; (3) if the trees die from below; (4) if dying white pines or dying chestnuts are affected with disease; (5) if death and discoloration of leaves is confined to one species only; (6) if the owner and if the state, allowing fires to rage, are guilty of contributory negligence; (7) if discoloration is caused by late frost, drought, or leaf fungi; (8) if the death rate within the smoke region is no greater than without, under the same conditions of geology and aspect; (9) if dying trees are normally covered with tree mosses, algae, and lichens; (10) if the death rate at the windward edge of a piece of forest is no greater than on the leeward edge; and (11) if the size of the annual rings of accretion is fairly normal.

To this outside work belonged the activities of C. A. Schenck and Company, in which some of my graduates engaged when field work was called for in connection with lawsuits, timber contracts, divisions among heirs, and so on. Most of the cases involved valuation surveys of the timber standing on a tract in the South. In the North, I had little or nothing to do, excepting some afforestations of a few hundred acres here and there, usually with white pines raised in the nurseries of my forest department. I do not mean to say that any of my graduate helpers were doing work with C. A. Schenck and Company which was not praiseworthy. Best of all co-operators were Thomas J. McDonald and Ralph G. Burton. The latter, unfortunately, contracted typhoid and died on one of our jobs. He was a daredevil of a man, unafraid of Southern swamps and Southern snakes and bursting with working energy; and he was a friend of all friends.

CHAPTER 15

The Parting
of Ways with Pinchot

SOME MONTHS AFTER President Roosevelt's visit to Biltmore it dawned on me that a timber canvass of the entire South was necessary — an investigation such as was begun twenty years later, in the years following the first World War, by the United States Forest Service. How could a federal forestry policy be framed by Roosevelt and Pinchot if the amount of timber available for the ax was not known? I wrote an article to that effect, outlining the scheme for the *Northwestern Lumberman.* In a subsequent number of that periodical, I was amazed to find an article written by Henry Gannett, chief geographer of the United States Geological Survey, which ridiculed my suggestion. Gannett, whom I knew personally and as a devoted friend of forestry, confessed to me in later correspondence that he had written his discrediting remarks at the instigation of Gifford Pinchot.

I had thought Pinchot incapable of such action. I had been co-operating with him in his Forestry Bureau; I had been asked to consult with him in Washington, personally several times, and sometimes unnecessarily, on conditions in Minnesota; I had written for him several reports on the possibilities of the pinelands in the South; and his first assistant, Overton Price, had been my pupil and was my devoted friend. True, I had criticized some of Pinchot's activities in my school lectures on forest policy, and I had advised my graduates to seek employment with the large owners of timberlands rather than with the Bureau of Forestry in Washington, because I wanted them to be foresters in the woods rather than foresters in office buildings. But I had never published, nor said in public, anything disparaging to him.

True, also, our attitudes with reference to the "lumber barons," as Roosevelt had styled them, were different. I wanted the bureau to co-operate with them instead of antagonizing and attacking them as enemies of the United States; for lumbering, in my opinion, was an essential part of forestry and an integral part of the studies and the lectures offered at any forest school. I remember as if it had happened today our heated arguments along this line when Pinchot and I, standing on the tracks of the Southern Railway in the Biltmore Nurseries, quarreled for an hour over the topic; and when he learned that in the school examinations at Biltmore a knowledge of logging and lumbering was weighed higher than that of silviculture or of any other branch of "scientific" forestry, Pinchot called me an antichrist. But I never considered these differences any reason for him to belittle me through a third person.

Another difference between us developed at that time. There was a movement then on foot in the United States Congress to establish some national parks or national forest reserves in the Appalachians, within states which, unlike New York and Pennsylvania, were financially unable to purchase from private individuals the lands required for state forest reserves. Most certainly the mountain forests had to be saved from destruction, and, since neither their owners nor the states were willing to preserve them, the nation should step into the breach. There had been very destructive floods in the streams draining the Appalachians; and physicians were urging that, in view of the growing statistics on tuberculosis in the eastern United States, fresh air resorts should be made available on a large and national scale. It seemed to me, as I considered the situation, that a multitude of small national parks would be useless in influencing the flood waters of the rivers and inadequate for the many millions of urban citizens requiring recreation. My proposition, published in an article entitled "The True Appalachian Park," was that all lands in the Appalachians above contour line two thousand feet were to comprise a national park — an area of approximately fifty million acres.[1] All owners of land within that area would be invited to co-operate with the Bureau of Forestry in the Department of Agriculture by pledging themselves (1) never to clear their woodlands entirely and never to burn them, but to assist in suppressing forest fires; (2) to permit the construction of national roads and trails and firebreaks on their properties; and (3) to cede the right of hunting and fishing on their properties to the people of the nation. Every owner taking the pledge would receive from the national treasury an annuity equal to the amount of his land taxes. Responsibility for fire prevention, road building and maintenance, enforcement of the state fish and game laws, and compliance of landowners with their pledges would be placed in a staff of forest rangers organized, as in the state forests of France or like the Feldjägers in Prussia, after the army pattern. I prophesied that every owner whose woodlands were made accessible for forest roads would be tempted to practice conservative forestry for the simple reason that it would be more remunerative for him than destructive forestry.

On the basis of my own experience in Pisgah Forest and the prices then prevailing, I submitted the following "budget" for the proposed park:

For the maintenance of a park of say 50,000,000 acres pro-

[1] A mimeographed copy of the article is in the possession of the American Forest History Foundation, Minnesota Historical Society, St. Paul.

tecting the waters, the public health and the lumber industries, keeping the mountains productive, acting as the largest game preserve in the world, etc., the U.S. would have to pay annually as follows:

1. Taxes reimbursed to woodowners $ 250,000
2. Salaries of 1,000 rangers at $600   600,000
3. Road-building, 500 miles per annum at $300   150,000
4. Trail-building, 1,000 miles per annum at $15   15,000
5. Superintendence by the Division of Forestry   35,000

            Total annual cost $1,050,000

On the credit page I made the following annual estimates:

1. Losses from inundations checked $ 10,000,000
2. Losses from forest fires avoided   2,000,000
3. Sickness at home decreased   60,000,000
4. Fifty million acres, producing annually $3 worth of finished commodities   150,000,000
5. National sport, hunting and fishing   20,000,000

            Total annual revenue $242,000,000

And here I added: "Even if this estimate is much exaggerated, 100 times exaggerated, even if protection of the lowlands, improved health conditions, national sport, lumber industry and proper use of mountain lands otherwise barren were worth to the people only $1,050,000 a year, we should be ready to spend that amount!" I admit that this queer sort of budgeting appears to me today absolutely puerile and fantastic. In the year 1900 I would have defended my figures in dead earnest. My proposal closed with the following:

The true Appalachian Park will stretch from the doors of Philadelphia to the gates of Atlanta, from smoky Pittsburgh to gay Chattanooga. Its lowest outskirts lie at an altitude of say 2000 feet, where agriculture recedes behind forestry. Its highest summit, Mt. Mitchell, kisses the clouds at 6700 feet elevation. This park, embracing the high mountainous sections of ten states, is the most useful National Park that ever existed. We have got to have it!

Gifford Pinchot did not share my views. I am afraid that nobody did at that time. I had been talking through my hat; I had had a pipe dream. And yet today, in the year 1952, I maintain that the United States, by neglecting this or similar propositions, has lost a wonderful opportunity to put forestry on the go in the Appalachians.

What times those were! For example, taxes at half a cent an acre a year; good rangers satisfied with a monthly salary of fifty dollars; wagon roads sixteen feet wide built at an average cost of three hundred dollars a mile and trails at fifteen dollars! The basal figures of my Appalachian budget were simply those costs which I had to reckon with in my Pisgah Forest before the boom of the tanning industry arrived in that blessed section of the United States.

Conservative forestry in those days was as much of a novelty in the United States as were the great American industries prior to the McKinley tariff. The American steel industry, the chemical industry, and the other great manufacturing industries were babies if they existed at all. For the manufacturing babies, the McKinley tariff acted as a mother; and its protective influence was so excellent that the babies developed within twenty years a capacity to defend themselves against any competition. But the McKinley tariff did more than protect. It invited and induced the capitalists of the United States and of the world to put their money in American factories by promising them good returns on their investments.

In 1900 conservative forestry was a very small baby in the United States; on private lands it was a mere embryo. What my Appalachian Park proposal amounted to was a financial inducement to landowners to conserve the productiveness of their woodlands. History proves that nowhere are private forests conserved when the destruction of the woods is more remunerative than their conservation, or when inducements towards conservation are not offered. Inducements may include reduced taxes, good roads built at public expense, feudal rights, tariffs on imported forest products, and so on. Pinchot's policy did not include inducements: in the West, the "lumber barons" were to be coerced into forest conservation, and in the East, as in the West, the compulsion system of forest conservation was to be applied through the creation of national forest reserves or national parks.

There was one exception. About the year 1905 Pinchot seems to have considered favorably the feudalistic idea of entail, that is, of transmitting forest property to specified heirs of the owner for a long period of years, with the proviso that the heirs practice or continue to practice conservative forestry. In the United States entail was forbidden; but there was a Massachusetts law which, in its effect, allowed an owner to settle lands or tenements on a certain line of his descendants for some forty or fifty years. The idea was good, but it was not put into practice; and it was not inducive enough for the needs of American forest conservation on private lands.

And today, what influences have the small areas of national forests and national parks in the Appalachians, or those of the entire East, had on floods, on public health, on public sport, and on lumber production? What influence have they had on the aims which, in the meantime, have led to the establishment by Congress of the Tennessee Valley Authority, which has spent millions upon millions of dollars on the watershed of just one river? I am confident that results like those obtained in the Tennessee Valley might have been obtained, on a scale of fifty million acres, under "inducements" akin to those mentioned in my Appalachian park proposal.

CHAPTER 16

A Portent
of Coming Events

In 1901 some leading men in Kentucky and Tennessee began to eye the need for forest conservation in their states. As a consequence, attempts were made to form state forestry associations. I was invited to assist at their birth. When I spoke in Louisville upon the invitation of the Commercial Club, however, the majority of my audience of six hundred persons consisted, not of businessmen, but largely of their beautiful wives and daughters who, I suppose, expected a flowery oration on the beauties, but not on the economics, of forestry. They could not be interested in a lecture based on the belief that forestry on a large scale was impossible in Kentucky unless it promised to be a remunerative investment. To them forestry at that time was a luxury on some beautiful estates, such as I had seen near Louisville on the Norton and Ballard places, where there were unforgettable tulip trees four feet through, standing in mixture with beech, a sight excelling any in Pisgah Forest. And never have I seen in Europe nor in the United States a better museum piece of a selection forest of beech than a stand on the bank of the Ohio River in one of these family estates.

In Nashville my audience was different. There were present professors of the University of Tennessee and the University of the South and J. H. Baird, editor of the *Southern Lumberman*. No need to say that neither there nor in Louisville was there any positive result of my addresses on state forestry. As a souvenir of the Nashville meeting, however, I cherish a book on "the philosophy of botany" written by the university professor of that subject and autographed for me.

O Lo! The philosophy of botany, the philosophy of the trees! Although I held a degree of doctor of philosophy, I ridiculed the idea of any botanical or dendrologic philosophy. Now, fifty years later, I have come to the conclusion that a philosopher cannot do better than to study religion, democracy, and economics in the light of the trees, nay, within the trees.

Religion! Listen to what Dr. Harry E. Fosdick has to say: "If a man is primarily growing a soul, he can capitalize anything that life does for him — as if he were one of the trees on the Maine coast:

> The Southwind warms them,
> The sunshine nourishes them,
> The northeast gales strengthen them,
> The winter cold toughens them.
> All weathers go to make a great tree."

And democracy! If you want to see democracy at its best, study it in the common tree standing by the wayside. Mark the co-operation of its leaves and its roots for the common good. Every leaf respects its neighbor. All work together while the sun shines. There are no strikes, no class hatred, no envy in lesser leaves of those occupying preferred positions; all are making buds in the spring to have the wherewithal ready for the next year. And this democracy is noiseless!

Economics! No effort made within the tree is wasted. The water supply of the tree is better regulated than that of any city in the United States. No architect can provide for better construction than the tree with the minimum of material used. Every cell formed — and there are billions of them formed annually — is at once placed in the very spot where it will serve best. Atom splitting, new with our scientists, is twenty million years old with the leaves of the most ordinary tree. And never are there any debts to be paid by the coming generations!

Upon my return from the Nashville expedition I found waiting for me at Biltmore one Captain George P. Ahern, who as a young army officer had explored and studied the forests of the Rocky Mountains and the Pacific Coast. He told me that the government was sending him to the Philippines to have charge of some fifty million acres of woodlands there and that he wanted to see what practical forestry looked like at Biltmore. O Lord! My kind of forestry was not applicable, I was sure, to the Philippines. If he desired a vision of his own possibilities, he should have visited the forests of Burma or those of Java, I thought. But I enjoyed his visit immensely. Here was a man with enthusiasm for a great task. He had had no education in forestry and almost none in botany, which is so important for the tropical forester; but he had devotion and he had the holy spirit of forestry. No more was needed for success, provided his health could withstand the climate of the tropics.

Another visitor welcomed at Biltmore at that time was Austin Cary, a graduate in 1887 of Bowdoin College in Maine and a man with wide experience in lumbering and enthusiasm for forestry. Apparently he had sufficient means to allow him to be a free lance and to buy some woodlands with a view of putting into practice what he believed in theory. I am sorry, indeed, that I did not have a chance in the following years to discuss with him the many problems in forestry on which we two agreed. We were both convinced that private forestry in primeval woods of the South and West was bound to begin with forest destruction; but that there was nothing to hinder it from rising from the ashes like the Egyptian phoenix except the lethargy of the legislatures, the apathy of the people, and the unsatisfactory prospects held up to forest owners.

The prospects of my Biltmore school did not improve during 1901. To increase attendance I sent out in October no fewer than two thousand circulars. The majority of the recipients were lumbermen. The theoretical and practical aspects of the Biltmore school were described: expenses, five to ten dollars weekly for board, six dollars monthly for the keep of horses, and two hundred dollars yearly for tuition; lectures to cover all branches of forestry and lumbering; daily excursions on horseback to give the student an intimate acquaintance with the practical and administrative side of forestry, which was defined "as the art of developing and of exploiting forest investments." As a special inducement for those enlisting for 1902, a three-months trip through the European forests at a cost of three hundred and fifty dollars was offered.

The circular did its work. I got a larger number of applicants than I had expected and than I had room for. Fortunately my new office and school building on the Biltmore Estate was about ready for occupancy. It was situated on the north bank of the Swannanoa River, and it contained a large classroom, a small anteroom, my office, and a small clerk's room. The salesman's office for fuel wood going into Asheville was at the gate of the estate a few feet distant. The woodworking plant was some two hundred feet to the west on the riverbank. Lo, I was in real clover! I had an office of my own. I was no longer compelled to work and lecture in McNamee's office in Biltmore village, with continual disturbances from the outside and the inside. George Vanderbilt's proclivity for building had helped me rather unexpectedly to a workshop of my own.

Biltmore village had been completed. It contained no

houses for the workers of the estate. All houses, which were built of the same material and of the same style, but no two actually alike, were for lease to whomever wanted to lease them. Many Ashevilleans removed to Biltmore because the rents were cheap and the houses were clean and new, all with baths and otherwise modern for the times. And there were clean macadamized streets and a parsonage close to Biltmore Church, which had a fine choir and a good organist, all paid for by Vanderbilt. There was a small hospital, too, open to those who wanted to use it.

In addition, a number of fine country houses had been built by Vanderbilt on Vernon Hill in Victoria, a suburb halfway between Asheville and Biltmore. All were completely furnished and ready to be occupied by wealthy persons wishing to spend a few summer or winter months in beautiful surroundings. For these the rents were rather high, and the majority of them were hard to lease. My department, leasing one of them off and on to acquaintances of mine, drew the real estate agent's fees of five per cent. Among these tenants were the F. W. W. Grahams of Philadelphia, the John L. Kusers of Bordentown, New Jersey, and Mr. and Mrs. Wilkes, Canadians. Wilkes, who was very fond of hunting, obtained a hunting lease, to the benefit of my department, for that side of the Biltmore Estate situated on the left bank of the French Broad River.

After erecting a dining house and a bathhouse at the Buckspring Lodge, and after spending two weeks in my humble cabin in the Pinkbeds, Vanderbilt told me that he was going to build another mountain lodge in the Pinkbeds or on Davidson River and also another mountain home for me. Before doing this, he wanted to spend a few days in the valley of Davidson River, but not in tents. There was no cabin for his party, and so I pledged myself to erect for them on the headwaters of the Davidson River a log cabin with six bedrooms, a dining room, a kitchen, servants' quarters, and a

shed for the mules and horses, all within three days. I knew
that I could do it because I had two road crews at work in
that vicinity; and I knew the speed with which my rangers
were erecting the cabins required for their crews of workmen.
I had only three days because the starting day for the trip of
Vanderbilt's party could not be postponed.

And thus was built "Three Day Camp" on a lovely grassy
spot — six rooms with a balcony in front, a porch in the center, and a dining house in the rear. One single red oak tree
furnished all the split lumber required for floors, sides, roofs,
and doors, the stump of the oak forming a huge table for teas
or breakfasts and whatnots. At this lodge I had one of my
most happy outings with the Vanderbilts. The favored guest
of the party was Mrs. Vanderbilt's brother, Roy Dresser, who
had just been made president of the new United States Shipbuilding Company. It was the first time that he had been to
Biltmore. The weather was fair. We had capital trout fishing
and some glorious excursions over new trails to Pisgah Ridge,
to the Balsam Mountains, and to the primeval woods of fir
(*Abies fraseri*) and spruce (*Picea rubens*) skirting the ridge.
And we climbed to the top of Lookingglass Rock on a series
of rustic ladders.

The Vanderbilts, fascinated with a part of the property
that they had never seen before, stayed on when Roy Dresser
had to leave. Going through my mail in a hurry on the last
day, I chanced by mistake to open a telegram reading: "Have
bought for you 100,000 shares. Roy Dresser." The telegram,
of course, was for Vanderbilt. I was ashamed to have opened
it, but there was no remedy but to confess to my oversight.
Little did I suspect that the telegram portended ominous consequences to me, to the Vanderbilts, and to the whole Biltmore enterprise — that it was a random cloud in otherwise
sunny skies.

At that time the rangers wore forest uniforms when a
Vanderbilt party was to be convoyed. From the various cuts

A High-wheeler in the Michigan Woods
Left, Mrs. Schenck; right, Mrs. Hermann von Schrenck

Ox Teams at Work in Pisgah Forest

Dr. Schenck in German Riding Costume,
about 1905

worn by German rangers, Vanderbilt had selected the style used by the state forest rangers of Saxony. The American rangers did not like to wear them. A uniformed ranger did not fit into the Appalachian wilderness; and when the fronts of the uniforms were soiled with chewing tobacco, as most of them were in a short time, their unsuitability was badly accentuated. Vanderbilt soon concluded that uniformed rangers were somewhat un-American, and no second set of uniforms was imported when the first set was used up.

Naturally, I was enchanted that Vanderbilt actually was becoming interested in Pisgah Forest, and I had high hopes that my dreams and plans would bear fruit in due course. To be sure, I had to sacrifice to Vanderbilt's pleasure trips more of my time than my other duties, and notably my duty as schoolmaster at the Biltmore Forest School, should have permitted. Franklin W. Reed, one of my best graduates, made things a little easier for me by acting as assistant forester in Biltmore when I was absent in the mountains with the Vanderbilts and with their parties who came to Biltmore in carload lots in Vanderbilt's private car.

There were about three hundred souls living in Pisgah Forest, some of them on farmlands leased to them, some in workmen's cabins built by me of heavy hewn timbers, and all of them loyal to Vanderbilt and to me, as far as I could judge. The public road leading along Davidson River had been abandoned as such and had been replaced by a private road on a good grade, with bridges spanning the former fords. Two portable mills were at work cutting selected trees for the lumberyard on the railroad at Pisgah Forest Station; and the deliveries of tanbark and chestnut wood were going forward in good order. I was having capital succcess in acquiring at cheap prices the interior holdings which studded Pisgah Forest and prevented the building of a system of private roads needed to protect the property from moonshiners, poachers, and fire fiends.

Naturally, the land prices of interior holdings would have risen from month to month as it became known that I was buying land for Vanderbilt. To prevent that rise I made it known that I had no more than three thousand dollars annually to spend on land acquisition. By this means I established competition among the owners, most of whom, off and on, wished to sell. When the three thousand dollars was exhausted, I gave my notes bearing six per cent interest due in the next or in the second or third year following. This, of course, was with Vanderbilt's consent. The banks in Asheville were willing to pay spot cash for these notes, which were considered good investments. Some of them I bought up myself. All in all, I spent about twenty thousand dollars while forester for Vanderbilt in acquiring for him interior holdings of Pisgah Forest.

During 1901 there was one event that saddened me. The German authorities in forestry had failed to renew my leave of absence, and I was informed officially that my name had been cancelled from the list of men connected with the German Forest Service. I was given, however, the rank and title of a German Forstmeister to sweeten, I suppose, the pill that was bitter for me. I was glad for this gesture of good will, nevertheless, because the rank of German Forstmeister helped me in my later migrations with my Biltmore students through the woods of Europe, and it made me feel in my innermost soul that I was working, though in America, also for my fatherland.

CHAPTER 17

Reverses at Biltmore

THE UNITED STATES SHIPBUILDING COMPANY continued in 1901 and 1902 to hold my attention and that of the American exchanges. The big bubble had burst; the stocks dropped rapidly in value, and Vanderbilt's building propensities came to an abrupt end. He had lost, by one single investment, a great part of his inherited fortune. Biltmore House became a white elephant. The nonpaying farms were closed. Charles McNamee left because there was no more need for a general manager of the Biltmore Estate. And there were no more Vanderbilt guest parties arriving from New York in the private car. The private car was sold. When I presented my next forestry budget for a working fund, Vanderbilt told me that I should do what other private parties were doing: I should borrow money at the banks and refund the loans out of the income to be derived in subsequent months.

Thus the year 1902 became the turning point in the affairs of all enterprises connected with the Biltmore Estate: theretofore expansion, thereafter curtailment. In that same year I was scheduled for another term in the German army, to be served during my biennial vacation as stipulated by my contract with Vanderbilt. After the sad experience of my last term served in 1900 I was most anxious to get rid of this service. I wanted to have some time free, during my German vacation, for my beloved parents, and time for the students accompanying me to Germany for trips through the woods of Europe. Fortunately, upon arriving in Darmstadt, I was told that my military duties had been dispensed with and that I was free from active service except in the case of war.

My absence from America in May and June, to my regret, prevented me from attending the annual meeting of the National Lumber Manufacturers' Association, where I had been asked to deliver an address and where I had hoped to meet the leading lumbermen of America. The invitation was renewed, however, and my desire was gratified, a year later at the meeting of that association in Minneapolis in August, 1903. From that time on the lumbermen were my friends. But what are friendships unless they result in deeds? My hopes that the leading lumbermen would send their sons, after their graduation from college, for a year to the Biltmore Forest School were not gratified. Naturally, these sons wanted to enter at once into their fathers' businesses to get practical experience rather than further doses of theoretical knowledge.

My appreciation of the lumbermen's situation respecting "scientific forestry" and my defense of them against the attacks of President Roosevelt and Gifford Pinchot sharpened Pinchot's hostility. In this spirit he wrote a letter to Vanderbilt asking him to close the Biltmore Forest School because its teachings did not conform to his views, because they were antagonistic to the development of forestry in the United States, and because they were producing a class of foresters derogatory to the graduates of the Yale Forest School. Vanderbilt showed me Pinchot's letter and asked me what his answer should be. It was easy for me to prove that the foresters from Biltmore had been placed by Pinchot in responsible positions in his own government service and that most of them had made good; that Pinchot himself, in previous letters, had praised their knowledge of forestry as shown by them in the federal competitive examinations. One of them, Daniel D. Bronson, had just ranked first among seventy-six competitors. I added that the help of the students in many lines of work on the Biltmore Estate and in Pisgah Forest was essential to him and to me. Vanderbilt sided with me, and Pinchot's attack was fought off.

It was only natural that after this experience my devotion to Pinchot came to an abrupt end. I thought it unfortunate that Pinchot's theoretical training at Nancy, the French school of forestry, caused him to identify the term forestry with the term silviculture. In the French idiom, forestry and silviculture are identical; in the English language, silviculture is the art of raising and tending a new generation of forest trees, and this art, although essential, is but a small part of forestry the world over.

The Vanderbilts spent the entire year of 1903 in Europe, evidently with a view to economies. Adding to their financial troubles, the stocks of the New York Central Railroad declined some thirty per cent in a few weeks. The great mansion, Biltmore House, with a monthly running expense of six thousand dollars, was closed in Vanderbilt's absence. My own expense budget was cut to less than fifty per cent of my recommendations, and the macadamizing of the main roads in Pisgah Forest was postponed for two years. Vanderbilt had undertaken too much in too many places. There were his other grand places — in Bar Harbor, New York, Paris, and Washington — to be maintained in a style befitting Vanderbilt traditions.

In a sense I was glad that economies were indicated and demanded. If forestry at Biltmore was to be a pattern for other private owners of woodlands in the United States, it was necessary for me to prove that it had been established without the wealth of the Vanderbilts.

At that time the number of winter visitors at Asheville had

risen to five thousand and that of summer visitors to more than ten thousand. The lines of the Hendersonville and Brevard Railroad, on which Pisgah Forest was depending, had been extended to Lake Toxaway, an artificial lake with beautiful beaches and with a large hotel on its shore. It was planned to extend the railroad to Georgia or into South Carolina, and this extension would make possible competitive freight rates from Pisgah Forest to the North and to the West. After long-continued efforts and several visits to Washington, I had succeeded in obtaining daily rural mail service for the folk living in Pisgah Forest. The preachers who came to the Pinkbeds for Sunday services were no longer detained by floods in the rivers, since all fords had been replaced by solid bridges to meet the requirements of the postal authorities. And there were two stores for general merchandise, both belonging to my forest department, one of them in the Pinkbeds and the other at the mouth of Avery's Creek in the valley of Davidson River. Civilization had definitely arrived, after a struggle of eight years! While there was depression within the Biltmore Estate, there was none in the rest of my little world.

On my Sundays in 1903 I was too busy to preach in the humble substitutes for churches in Pisgah Forest. My last preachment, in the Pinkbeds, had been a failure anyhow. I had taken for a text the fourth chapter of Saint Matthew, and I spoke of the devil, claiming that there is no exterior devil, but that the tempter lives within ourselves in the form of lust or fondness for liquor, for the other man's wife, for loafing, for lying, and so on. When I had finished, my best ranger, old Jimmy Case, stood up and said: "Dear brethren, this is the last time I shall be praying with you. I cannot live any longer with Dr. Schenck who does not believe in the veracity of the Bible. There is a living devil and I can prove it from one hundred sayings of the Holy Book."

Thereupon he proceeded to read a number of Biblical

passages referring to a bodily devil. There was a great commotion. I had obtained a result entirely different from that intended and had shocked the best element among my cooperators. There was nothing left for me but to confess openly that I had erred, that Jimmy Case knew the Bible better than I, and that I would cease to preach to the good folk so as not to mislead them by any further errors of mine. And so my ministerial work in the Pinkbeds came to an end with the devil at its tail. From that time on I was nothing but a listening member of the congregation, and sometimes, in the Pinkbeds, their organist.

In August, 1903, in Minneapolis I attended for the first time the annual meeting of the National Lumber Manufacturers' Association, and delivered an address on forestry. I was a member and, for some time, also a director of the Hardwood Manufacturers' Association, which formed one of the many sections of the national association. The Hardwood Manufacturers' Association was at war continually with the Hardwood Dealers' Association. The grade inspection rules of hardwood lumber were the main point of difference, those of the manufacturers being more lenient than those of the dealers. Both associations were maintaining staffs of official lumber inspectors, who were brought into action whenever there was a dispute about quality between shipper and customer.

While I was a member of the hardwood manufacturers I could sell no lumber according to manufacturers' inspection because my customers were dealers. Since the output of my mills was too small to sell directly to lumber consumers, there was no chance for me to sell at all unless I sold through the dealers. The troubles of the hardwood manufacturer result from the great variety of his output. I had in the woods and in my lumberyard no fewer than twelve species of lumber — yellow poplar, red oak, white oak, basswood, soft maple, hard maple, yellow birch, black birch, black gum, chestnut, hickory, and hemlock — each species selling in some six to eight different grades. Thus it was hard for me to sell a straight carload of lumber of one grade, thickness, and species. Cucumber sold as yellow poplar; chestnut oak was mixed with white oak; some black oak was smuggled in with red oak. It is easily seen that a hardwood sawmill has a difficult task. Its sawyer, its edgerman, its trimmer man, and its off-bearers *must* know the inspection rules of many species of trees and must know them well if the highest value is to be obtained from every log and every board sawed from it.

I tried hard all my life long to learn and to know those hellish inspection rules of lumber and to instruct my men at the mills, giving special weight in competitive examinations to the best man; and yet I am conscious that much money was lost at my mills because of ignorance of lumber inspection rules. The removal of a small knot or a small strip of sap at the edge of a board may decrease the size of the board by ten per cent, but may increase its value by twenty per cent. Thus, aside from the inspection rules, the relative values of the grades of lumber from twelve species should have been known by every employee in the sawmill.

No wonder that most foresters, despairing of the possibility of learning these intricacies, deny the forester's necessity for any knowledge of lumber and lumbering. They are satisfied to do a little pruning in their woods to prevent their trees from being too knotty, and to remove crooked trees by thinnings to minimize the number of low-quality boards coming from crooked logs. But what do the foresters know of the differences in the value of boards, otherwise alike, the one sixteen feet and the other ten or twelve feet long? How can a forester see to it that the bole of a felled tree is properly dissected unless he knows the rules of lumber inspection? How can he appreciate the necessity of preventing the log ends from splitting unless he knows these rules? And what about the influence of the ambrosia beetles, the blue-stain fungi, and other insects and diseases on the value of a log? It is just impossible for him to give an answer unless he knows the inspection rules of lumber.

The students of the Biltmore Forest School were forced continually to enter into the secrets of lumber inspection. I had set aside some stacks of lumber, all species and all grades mixed in them and every board numbered. The students were required to inspect these boards, number by number, and give me the species and the lumber grade of each board. It was but natural that almost daily in my lectures I touched on the conditions of the lumber market, the demand for lumber, the prices obtained, and so on. In a word, the Biltmore students were meant to be lumbermen-foresters. I could not help being as proud as a peacock when some Yale students, dropping in at some of my lectures, assured me that never at Yale had they had any lectures as interesting as mine.

George Vanderbilt's finances, instead of improving, continued to grow worse; it was reported that he had lost additional sums of money. As a consequence, Edward J. Harding, financial controller at the Biltmore office, and his staff were dismissed and the chief of the farm department, my friend George Weston, was forced to resign because he was unable to make ends meet. I cancelled my trip to Germany, planned for 1904, because I could not leave the ship in a stormy sea of financial uncertainties. Road building and land purchases were discontinued. I made no improvements in 1903 except to add a hatchery for English pheasants in the Davidson River Valley with a view to increasing the income of Pisgah Forest from hunting.

In an attempt to raise money for his Biltmore Hospital, Vanderbilt opened Biltmore House to the paying public of Asheville with a concert given by the famous singer, Johanna Gadsky. But alas! The receipts expected were out of proportion to the effort. The audience consisted of only five hundred

paying guests, some of whom complained that they could not
see the singer and that they were not seen themselves, dis-
tributed as they were in several rooms of Biltmore House.
Vanderbilt did not repeat the attempt.

During this period we had guests aplenty at my house,
often too many for Mrs. Schenck to accommodate. This con-
flux of guests under my own roof was a continual joy and
inspiration. It is so much easier to lecture interestingly when
the mind is well fed by intelligent visitors. Among these
guests were a zoologist who could imitate the voices of all
birds without any instrument; some English-Indian foresters
who had been under my tutelage when I was assistant to Sir
William Schlich on his European forestry tours; Dr. Hermann
von Schrenk of the United States Bureau of Plant Industry,
famous for his knowledge of mycology of trees; Dr. Charles
H. Herty, a chemist who later became widely known for his
work in the utilization of Southern pines for papermaking;
and Dr. William A. Murrill, the famous botanist. Believing
that any essay read is more interesting to the listener if he
knows its author personally, I invariably asked learned visitors to lecture to my students assembled in the classroom.
Always I was eager for my students to obtain what a university rarely then gave — a knowledge of men as well as a
knowledge of forestry.

Opportunity to meet American leaders was provided also
by meetings in Washington, D.C., of the Society of American
Foresters, founded by Gifford Pinchot in the fall of 1900.
Often, in connection with these meetings, there were recep-
tions at one or the other of the great houses in Washington.
Thus I chanced to be present in the home of Alexander Graham Bell, inventor of the first telephone, when he explained
the accident that led to his discovery. Impelled by the fact
that Mrs. Bell was deaf, he had set out to construct, not a
telephone, but an instrument to facilitate hearing. The original first telephone was shown. I was deeply impressed by

the modesty and simplicity of this man, who seemed to be unconscious of his accomplishments and fame.

The Society of American Foresters in those days embraced fifteen members, four of them Biltmoreans: Overton Price, Edwin M. Griffith, Frederick E. Olmsted, and myself. I confess I did not take much stock in it, believing we were too few in numbers and too ignorant of forestry to be representative; nor was I sympathetic with the term "scientific forester" so often heard in the United States, holding as I did that a forester is a practitioner like a physician or a lawyer. Pinchot differed from me; he was eager for his foresters to obtain a high standing in the eyes of the public and of Congress, and to that end to become allied with the scientific organizations having headquarters in Washington. He was right and I was wrong; but I have felt that the stress he laid on forestry as a science was instrumental in deepening and broadening the abyss between forestry and lumbering in the United States.

In October, 1904, Mrs. Schenck, who had had to take her vacation alone in Germany, returned accompanied by my youngest sister and my mother, who was beloved by everybody.

In February and March, 1905, I attended some Canadian forestry meetings in Quebec and Montreal. What a beautiful city Montreal was! How interesting were the winter sports on Mount Royal — sports I had never before seen. At the meetings I made the acquaintance of Sir E. G. Joly de Lotbinière, owner of one of the old French-Canadian seigniories near Quebec, and interested in forestry. He told me that his father, Sir Henri Joly de Lotbinière, recently lieutenant governor of British Columbia, had been practicing forestry on his seigniory and had been trying to introduce it into British Columbia; and he was eager for me to see this seigniory. Together we traveled from Montreal to Quebec, a glorious trip through ice and snow, the tide in the St. Lawrence driving whole mountains of ice upstream. From Quebec we proceeded by rail and sleigh to the Lotbinière woods, where there was a snug hunting lodge with all the comforts required for our stay. We inspected the cutting of pulpwood and the arrangements for driving it down the small and the larger streams; we saw the ash trees dying from causes unknown; we saw other hardwoods, all worthless at the time; and we saw the gigantic stumps of the white pine, five feet high and two and more feet through, which Grandfather Lotbinière had felled and sold at a time when white pine alone had any value. We looked in vain for any second growth of white pine. There was none excepting one seedling standing on the stump of one of the old trees. Otherwise the species was extinct. I came to the conclusion that a student of forestry can learn more of nature's silvicultural secrets in the primeval woods than in the cultured woods, and most certainly more than in the classroom.

President Roosevelt in the meantime was advocating forestry in many interviews and public addresses and in some presidential messages to Congress. In 1903 he had appointed a Public Lands Commission of three members charged with making a report on the federal land laws. Roosevelt knew from his own Western experience that these laws were being used and misused for thefts of valuable timber and mineral and pasture rights. To all misdeeds and frauds, so it was rumored, the Department of the Interior, in charge of the public lands and responsible for their disposition, kept its eyes and its ears shut. There were the Homestead Act of 1862, the Timber and Stone Act of 1878, the Desert Land Act of 1877, the swamp land acts, the railroad grant acts, the mining laws, and the octopus of the Indian land laws. Of course no report on the condition of the public lands could stop the prevailing abuses, but President Roosevelt succeeded in drawing attention to the incompetence of the Department of the Interior and to the advisability of transferring the forest reserves from that department to the custody of the Department of Agriculture, where the government's Bureau of Forestry was located.

By the Philippine Forest Act of 1904 the use of the forests in the Philippine Islands was regulated, but the Philippine Bureau of Forestry was left in the control of the Department of the Interior. In the Hawaiian Islands a Division of Forestry, headed by Ralph S. Hosmer, a graduate in Yale's first class of forestry in 1902, was created at about the same time, and some 250,000 acres of forest reserves were established. Most important, however, was the activity in the individual states of the Union. The legislatures were becoming aroused by the continual propaganda by Roosevelt and Pinchot; chairs of forestry or forestry schools were being established or considered by state universities; state foresters and state boards and departments of forestry were on the increase; improved forest fire laws were being enacted; state forestry associations were forming; and a few states, those financially able to do so, were establishing state forest reserves. Thus the five years following the assassination of President McKinley gave stimulus to forestry everywhere.

I did not take a hand in this movement. As a matter of fact, I was not so deeply interested in this field as I might have been, because I could see nothing in all those laws to ameliorate the conditions under which a private forest owner was forced to operate when he wanted to practice conservative forestry. To me it seemed as if the main issue was merely clouded by these state and federal activities, since nothing was being done to make forestry on private lands a safe and remunerative investment. It appeared to me that this method of pushing public forestry to the fore and private forestry to the rear was unwise and undemocratic.

CHAPTER 18

A Visit
from Peter Thomson

PUBLIC PROPAGANDA FOR FORESTRY in this period reached its
climax in 1905, when President Roosevelt addressed an
American Forest Congress held in Washington under the
auspices of the American Forestry Association. All the important, and many not so important, persons interested in
forestry attended, including foreign ambassadors, leading
landowning lumbermen, and state and federal officials. Roosevelt made the main address, and with characteristic vigor he
attacked the lumbermen, calling them public enemies and
causing some to leave the meeting in resentment. I was
present, but I spoke only briefly on the conditions in North
Carolina and on the impediments to private forestry as a
business.[1] Here is what I said:

My connection with forestry in western North Carolina is
of a three-fold character: I am a lumberman, a forester and
a teacher.

I am a lumberman, and I must confess to being somewhat
afraid as a lumberman to appear before this audience. Still,
while in charge of a large forest in western North Carolina,

[1] As reported by members of the audience, Roosevelt discarded his prepared speech for an extemporaneous attack on the lumbermen. The prepared
speech, as well as Schenck's talk, appeared in American Forestry Association,
Proceedings of the American Forest Congress, 3–12, 437–439 (Washington,
1905).

142

I cannot help being a lumberman. Without lumbering no cash dividend is obtainable from forest investments. Therefore, I cut the trees, though . . . I do not cut all the trees — for the reason that it pays better not to cut all of them, under the conditions now prevailing in western North Carolina.

We are just beginning a new year, and, as new year's wishes are in order, I wish that every one of you were possessed of 50,000 acres of hardwood lands in the Appalachian range! If *you* were the owners of such timber tracts in our mountains, or anywhere in the East, *what* would you do with the timber? I ask your conscience, would you let the timber stand, or would you convert that timber, all of it or part of it, into money? We are in the habit of blaming the other fellow for cutting the trees. Now, pardon me when I ask: What would you do with the trees if you owned them?

Secondly, I am a forester, and as a forester I am meant to raise trees, partly by planting, partly by lending Nature a helping hand. The owner of the Biltmore estate, without doubt, would authorize me to practice more silviculture if he could consider silviculture (the raising and tending of a second growth) a remunerative investment; I had better, perhaps, say a safe and remunerative investment.

However, as fires annually rage over large sections of our grounds, it is hazardous — nay! it is almost folly — to invest money in silvicultural pursuits. At Biltmore we are forced to restrict reforestation to such regions in the proximity of Biltmore House in which we can control fires absolutely. In a large primeval tract covering 120,000 acres of backwoods, absolute fire protection is out of the question. Here I do not attempt to enforce regeneration, simply I allow nature to do the work as best she can, trying at the same time to protect the second growth from fire wherever it appears.

Foresters are very frequently, I think, of the opinion that the little trees — second growth — are really the best money makers. Foresters working in the Appalachians might just as well begin to change their minds. The fact has been pointed out to-day repeatedly that the price of hardwood stumpage is increasing rapidly. If that is true, the big tree is the best money maker, and really mature trees do not exist — moribunds excepted — where and as long as the price of stumpage advances rapidly.

In 1896 I sold many a fine white oak at fifty cents per thousand feet, board measure. I wish I could replace these trees. I would gladly put them back into the woods at $4 a thousand — because they are worth now $5 a thousand. In 1898 I got for similar trees $1.25 a thousand feet, board measure; in 1902 I received $2.50, and last year I found a man who was willing to give me as much as $8 per thousand!

Thus it happens that the big trees — the three, four, five and six-footers — are my pride, more so than the seedlings and saplings. I hold the big giants dearly; I refrain from cutting them, merely for the reason that they are my best money makers, the best part of my investments — and also the safest part of my investments since they are not subject to destructive forest fires. So much for the forest.

Finally, I am the director of the Biltmore Forest School, established at Biltmore, North Carolina, in 1898. I am delighted, though it makes me feel old, to see so many of my former pupils present in this hall. Permit me to use this chance for reminding them forcibly of my old demands and unceasing teachings — so often repeated with the regularity of a canary bird or of a whippoorwill — keep constantly before your eyes the fact that forestry *subserves* lumbering, that forestry *is* lumbering to a very large extent.

Silviculture and lumbering together will, I think, compose the work of the forester in this country for many a year to come. The greater portion of practical wood's work will lie in the line of lumbering, and the lesser part will consist of silviculture merely because silviculture is not as safe an investment at present, nor is it as remunerative as lumbering.

The time will come when the reverse will be the case. It will come when the superiority of conservative lumbering over destructive lumbering is clearly evidenced by the larger number of dollars which conservative lumbering can draw as a dividend from the forest.

Following the example set by Theodore Roosevelt, the governor general of Canada, Earl Grey, convened a Canadian Forestry Congress in Ottawa on January 10–12, 1906. Pinchot and I were guests of the governor at Government House. We were received in the family circle of the Earl, which included the young Lady Victoria Grey and the young Duchess of St.

Lidgerwood Logging Machine

McGiffert Loader with Running Wheels Raised

Roosevelt and Pinchot on a Trip down the Mississippi in 1907

Albans, as if we were old friends and as if the house were thoroughly democratic. The style at the daily functions of life, however, was that of the Court of St. James. It was the governor general, and not the ladies, who was first to enter the dining room, and when the meal was over he was first to leave, the ladies, including his wife, arranged at the right and at the left of the door and curtsying as deeply as though he were His British Majesty himself.

The Canadian Forestry Congress was opened in the presence of the governor general but without a message by him. Kings do not speak in public. Among the leaders invited to attend the congress were the high officials of the dominion government, the provincial governors, the leading holders of timber limits, the presidents of the wood-consuming industries, and the magnates of the Canadian railroads. Most of these men were given a chance to make themselves heard. The congress was presided over by Sir Wilfred Laurier, prime minister of the dominion, who made a deep impression on me by his eloquence and his address.

After the sessions were over, Earl Grey, who had expressed a wish to see some typical lumbering, his family, and I were loaded into the earl's private car and were switched during the night to the holdings of John R. Booth on the Ottawa River, where, so we were told, there were the finest stands of virgin white pine in all Canada. We were shown the cutting and swamping, the skidding and sledding, the loading and dispatching of a few cars of logs. Old John Booth, white pine king of Canada, was a plain, kind man, more familiar with lumberjacks than with the aristocracy of England. In the northern pineries I was impressed more deeply by the luxurious regeneration of white pine after fires than by the size of the mother trees, none of them measuring more than forty inches in diameter. I had expected trees as big as the tulip trees in Pisgah Forest.

So much for the meeting in Ottawa. As far as I could see,

its practical effect was none. In the province of Ontario, forestry was banished from the crownlands by the transfer of the Bureau of Forestry from the Interior Department to that of Agriculture. On the other hand, the University of Toronto established a department of forestry in 1907 under Dr. Bernhard E. Fernow. Prairie planting, at the same time, was greatly promoted by the appointment of Norman Ross, one of my best Biltmore graduates, as chief of the Dominion Forest Nurseries in Indian Head, Saskatchewan. An indirect effect of the Ottawa meeting may have been that much English and American capital was drawn to investments in timber limits, notably so in British Columbia, Quebec, and Ontario.

In the meantime important movements were on foot at Biltmore. The sport feature of Pisgah Forest, now that interior holdings had been bought up and the public roads had been closed, had become a valuable asset. I had succeeded in getting two committees interested in a lease of the exclusive hunting and fishing rights in Pisgah Forest. One committee was headed by Edgar B. Moore, proprietor of the Kenilworth Inn near Asheville. The other consisted of some leading New York financiers. Vanderbilt was willing at the time to forego the hunting and fishing rights on the Biltmore Estate proper, reserving for his personal friends merely a thousand acres immediately surrounding the mansion. He suggested as an additional inducement to surrender to the members of a proposed sport club his Brick Farm House for a clubhouse and all the grounds on the upper bottoms of the French Broad River for a golf course, against a corresponding payment in addition to the annuity of $14,550 which the hunting and fishing rights were to bring to my forest department.

Vanderbilt must have been in bad straits at the time. He was anxious for the club to pay the first ten annuities in advance and in a lump sum, an arrangement which did not suit me since I wanted annual receipts. The New York committee was headed by United States Senator John F. Dryden, president of the Prudential Life Insurance Company, and among its members were such well-known men as Franklin Murphy, the governor of New Jersey, and John I. Waterbury, president of the Manhattan Trust Company. I had several meetings with the New Yorkers, but after I had exchanged many letters with them and with Vanderbilt, and after much delay due largely to Vanderbilt's continuous absence from America, both deals fell through. I am sure the New Yorkers must have known at the time that the severe depression of 1907 was in the air and that the time was ripe for contraction rather than expansion.

All this time I was bound to the bridge of my ship stranded in a dearth of cash. Repeatedly I was forced to give my own notes to the banks for the benefit of my payrolls, and when the banks hesitated to accept them in larger amounts, Vanderbilt undertook to guarantee by letter their payment in due time. It was not much fun in those days to be the forester of the Biltmore Estate. In truth, there is not much fun in a forester's life, unless he draws the fun from nature's unlimited supply. When it rains, there is no more danger from fires, and the plantations thrive. When it shines, the roads are dry, the lumber dries fast, and the tanbark gets through. When it is cold, the market for fuel wood is buoyant, with no competition. And when it is warm, the logging operations in the woods are in full swing, and the work on the roads is no longer handicapped.

Thus, you see, the forester is in luck in all kinds of weather. And there are things or weights which tend to counterbalance his financial worries and disappointments, such as a spider-web covered with drops of dew in the early morning grass; eight tiny dachshund children, born to Minnie, his pet bitch; a garden show in Pisgah Forest, when good old Mrs. Duckett is awarded the first prize, a sewing machine worth $11.95, and weeps for joy; a thunderstorm beneath his feet viewed from

Frying Pan Mountain; a forest fire stopped singlehanded before it has gained any headway; and so on and so on.

In the meantime another project had come up which was of vital interest to me, for if it carried through it would provide a market for such pulpwood as was to be found in Pisgah Forest. The local newspapers were filled with announcements that a large wood pulp factory was to be built in Canton, Haywood County, immediately northwest of Mount Pisgah, by Peter G. Thomson, owner of the Champion Coated Paper Company in Hamilton, Ohio. Upon his invitation I visited him at his Ohio plant and was deeply impressed by his personality. Followed by two good-looking sons and a son-in-law, Reuben B. Robertson, he entered the factory, where I was waiting to meet him, at the stroke of eight o'clock. As he showed me through the plant, I could not help admiring a huge sheet of wet paper, some fifty feet long, floating between blasts of hot air as it was sprinkled with a fine crust of chalk, which in turn was pressed into the sheet from above and below by a series of heated cylinders.

Thomson was then buying the wood pulp required for the paper sheets, but in order to be sure of a steady supply of raw material he wanted to become a manufacturer of wood pulp. It was he who had bought up at ten dollars an acre those ten thousand acres on Pigeon River mentioned previously, which I had not had the vision to buy myself. Along with the spruce and Fraser fir on the tract, which were to be made into pulp by the conventional sulphite process, he proposed to convert chestnut wood chips to pulp by the conventional soda process, after first having extracted the brown tannic acid by a patented process which he controlled. For the conveyance of wood from the forest to Canton, a distance of fifteen miles, a huge flume was to be built. At the upper inlet of the flume a snug village with a church and a school was planned. The whole scheme was the most gigantic enterprise which western North Carolina had seen.

On my invitation Thomson and his wife came to the Pinkbeds, my headquarters in Pisgah Forest, and stayed with me
for a number of days. There I had a chance to learn all about
his plans, and I was overwhelmed by their magnitude. What
a pygmy I was in comparison! And what a pity that Pisgah
Forest did not contain any spruce or Fraser fir! Chestnut only
I had to offer, and that was contracted for delivery over a
period of years to the Adams' tannic acid plant at Pisgah
Forest Station. Thomson was not interested at that time in a
second growth of spruce or chestnut after removal of the
first growth, nor was he interested in forestry; and I was
unable to coax him to a changed attitude.

What I admired most in Thomson was his courage. Here
was a man such as is described in Kipling's poem "If" — a
man making one heap of all his winnings and risking them
on one throw of pitch and toss in a venture of really great
magnitude and on ground unfamiliar to him. And all was to
be done in keeping with prearranged plans. He was a man of
teeming energies, a leader in economics, bristling with new
ideas like that of using chestnut for the manufacture of paper,
or like that of fluming all wood required for his cellulose
works over a distance of fifteen miles. George Vanderbilt's
enterprise had been great, too, but how different was its
grandeur from Thomson's venture! In the last analysis, Vanderbilt's undertaking was a rich man's hobby; it had no
grounding in economics, and it proceeded haphazardly without any preconceived plans. No wonder that it was destined
to fail in the end.

I told Thomson, while he was a guest at my camp in the
Pinkbeds, that I doubted that the volume of spruce wood he
would require could be obtained in western North Carolina
for longer than ten years. I also doubted the wisdom of flumes
as conveyors, because of their dependence on a steady water
supply and because a flume works only downhill, while the
transport of the many goods required in logging operations

must go uphill. As it turned out, Thomson found soon thereafter that a flume as projected by him was incapable of carrying daily the three hundred cords of wood that his plant required. He found also that the supply of spruce was insufficient and that he had to mix it with North Carolina pine. In 1908 or so my firm, C. A. Schenck and Company, was engaged by Thomson to make a canvass for pine all over western North Carolina. In the meantime he had built the snug village of Sunburst, at the head of the West Fork of Pigeon River.

My connection with Reuben B. Robertson, Thomson's son-in-law and general manager of his North Carolina interests, was very close and very friendly and it has continued so to the present day. It was Robertson who took up forestry in later years on all holdings of the Thomson concern, and I hope that he has made a success of it.

CHAPTER 19

"Try to Sell Pisgah Forest"

THE BILTMORE FOREST SCHOOL, in those otherwise disheartening days, began to flourish, although it continued to be a one-horse affair. It had no donors to give it $150,000, such as Yale Forest School had received in 1905; it had no Weyerhaeusers to aid in establishing a chair of lumbering as they did at Yale. I was a little hurt when I heard of these gifts, and a little envious, although I would not have known what to do with a gift of $150,000 for my school or with a fund creating a chair of lumbering. Indeed, never in the entire time of its existence did the Biltmore Forest School have any financial or any personal assistance from anybody. No wonder that it was short-lived! But I have loved it dearly; and it continues to live in my memory cherished as much as if it had been my only child.

In 1904 Clifton D. Howe, a graduate of the University of Vermont who had received his Ph.D. degree in ecology and physiology at the University of Chicago, became my assistant forester, and a capital assistant he was — industrious, gentlemanly, absolutely reliable, intelligent, and, most important for me, a favorite among the students. No wonder that, when the province of Ontario, in 1919, wanted a successor to Dr. Fernow as dean of the Faculty of Forestry at the University

of Toronto, Dr. Howe was chosen for the position. He filled
the chair with remarkable ability for many years.

Among the teachers not already mentioned who came to
Biltmore during these years were Dr. Harry D. Oberholser of
the United States Biological Survey, who lectured every summer for four weeks on forest zoology; Dr. A. D. Hopkins and
C. D. Condon of the United States Bureau of Entomology;
Franklin Sherman, state entomologist, and C. S. Brimley,
state zoologist of North Carolina; H. O. Allison, professor of
animal husbandry at the University of Missouri; J. Girvin
Peters of the United States Forest Service; and Edgar D.
Broadhurst, a lawyer of Greensboro, North Carolina. These
auxiliary lecturers, by their close contact with the students,
by living with them in the woods, by daily excursions with
them, imparted to them a unique knowledge in subjects such
as zoology, astronomy, geology, mycology, and climatology.

The Biltmore students had built, mostly with their own
hands, a little clubhouse in close proximity to Biltmore village. It had a large general room, a small kitchen, and in the
loft two small bedrooms for guests, notably for graduates
coming to Biltmore on visits. By means of this clubhouse I
hoped to concentrate the drinking propensities of the students and to keep them away from the bars in Asheville. The
regular sängerfests were celebrated once a month from that
time on in the clubhouse.

Pardon me, dear reader, when I speak of the use of alco-
holic beverages in connection with forestry, for the two are
sometimes closely connected. In 1907 there was a movement
on foot in North Carolina to suppress all bars and alehouses.
Although I confess to being fond of alcohol in moderation, I
moved heaven and earth to lead this movement to success. My
workmen all too often were intoxicated; my mule teams were
left unfed in the stables from Saturday to Monday while the
muleteers were on drinking sprees; some of my students were
frequenting the Asheville bars too often. For western North

Carolina the advent of prosperity had come rather suddenly, and prosperity made it possible for the poor to buy alcohol more freely.

In Germany conditions were and are different. The Germans are actually raised, not on strong drinks, but on wine and beer. When my four brothers and I were as tall as the dining table we got a little sip from the daily wine bottle, which it was the custom of my father to empty after the evening meal. And when the annual big hogshead of Rhine wine arrived in Darmstadt and in our cellar there was great rejoicing in the family. I do not believe that any one of us German boys, by this kind of education, was made a drunkard. On the contrary, from my earliest days I was taught to enjoy a drink and to detest drunkenness.

In North Carolina prohibition had become a necessity. I was jubilant when, on May 26, 1908, the state was voted dry. Conditions among the workmen improved when the bars were closed. Among the upper classes conditions remained as they had been. The members of the fashionable clubs had their lockers at the clubhouse, where they kept their stores of liquor. And the wealthy could import from a wet state or from Europe whatever alcoholics they wanted, since no law of North Carolina could stop interstate traffic.

The establishment of prohibition in North Carolina was preceded by the arrival of that queer depression which fell from the skies in 1907. Some of the businessmen in New York City anticipated it, among them a rich tanner, Hans Rees, who predicted, to my amazement, that the high prices for tanbark prevailing in 1906 would not be maintainable in 1907. As I talked to him I thought, of course, that he was merely trying to close his contracts for tanbark deliveries at a reduced price. It proved otherwise. My stores of dry lumber suddenly became unsalable. My lumber customers failed to meet their notes for lumber purchased when they were due. Among them was Fred Brenner in Norfolk, my agent for lumber shipments to Europe. I was unable to obtain any cash in Norfolk. The banks there declared that the Brenner notes were good if I would only be patient. Patience! But there were payrolls and taxes due in my forestry department and they could not wait. Never was I in greater despair. George Vanderbilt was absent in Europe; but, had he been in North Carolina, even he would have been unable to help.

There was another sad occurrence for me in 1907. My paternal friend and protector and paragon, Sir Dietrich Brandis, died on June 28, 1907. I had seen him last in 1906, when I called on him in Kew Gardens while passing through London on the way to Darmstadt. He did not then seem to have aged. As usual, he was full of praise for the achievements of others — of Schlich, of Ribbentrop, of Pinchot. I have never met a man who had more reason to speak of his own achievements but who spoke less of them, giving all glory to his assistants and to his successors. His memory has continued with me as active, I believe, as it was in 1907. I have always tried to follow his example by working for the cause and never for myself. Immortality! What is the sense of it unless the impression that the living man has made continues for decades after his death?

The depression of 1907 was a terrible blow to my hopes; forestry at Biltmore seemed to be doomed. All work on the Biltmore Estate was stopped, and there appeared no need for me to stay there. I might just as well spend a few weeks in Germany. Therefore I took ship for Europe and spent Christmas with my mother. Then, accompanied by six of my students, I went to Vienna, applied for permission to inspect the forestry work of the Austrians in the primeval woods of the Carpathians, and had the most instructive forestry excursion of my life, thanks to the kindness of the Austrian authorities, who ordered their Carpathian chief, Oberforstmeister Hirsch, to accompany us.

Upon my return to Biltmore on February 25, 1908, I was

welcomed by a cavalcade of my forest students, who accompanied my carriage to my home on the Biltmore Estate. They gave me a wooden tablet on which, in memory of Punch, my pet riding horse who had died while I was away, were inscribed these words:

> Who is the man on a horse named Punch
> Riding along at the head of the bunch,
> Giving no time to eat our lunch?

The school had done well during my absence, but the lumber trade was still in the claws of the devil. The year was a presidential one; the elections were coming and the uncertainty of the future continued to hold the lumber market down. My accounts were deeply in the red; my department's debts with the banks were increasing. George Vanderbilt was very nervous, and Mrs. Vanderbilt was in tears whenever I called on her. Vanderbilt had ordered all his saddle and carriage horses to be sold and the grooms discharged. There were no visitors. I had the feeling that somebody, during my absence, had succeeded in undermining Vanderbilt's confidence in me and, what was worse, in the outcome of his and my enterprise in forestry. He told me that he was done with forestry and that someone else might now take up where he was willing to quit. "Try to sell Pisgah Forest for me," he told me. "I will pay you the usual agent's commission on the purchase price."

I did not know at the time whether or not he spoke in earnest. Would not his huge mansion continue to be a "white elephant" if it was deprived of Pisgah Forest as its hinterland and mainstay in the years to come? And was not he, like many aristocratic forest owners in Germany before him, dependent upon the woods to furnish income when it was badly needed? But apparently Vanderbilt was as short of cash as was his forest department.

As for me, I was threatened with dismissal, with a collapse

of all I had been working for during the best and most active years of my life, threatened with becoming a laughingstock instead of setting a sound pattern of forestry for the United States. All my forestry, all my teaching, all my ambition, all my anxieties were shipwrecked if Pisgah Forest were sold. And what was an agent's fee to me as a recompense, even if it amounted to five per cent of a million dollars? I had never worked for money alone. I did not make the slightest attempt to find a purchaser for Pisgah Forest. With the sale of Pisgah Forest, my beloved Biltmore Forest School would lose its working field, its demonstration field, its experiment stations, and its very basis. Vanderbilt, after giving me a hint of his intentions that one time, never returned to the topic when I met him, and I began to think the clouds had passed. In July the Vanderbilts left for France, stating that they would be absent for several months.

Before they left I had a surprise. Lady Alice Grey, wife of the governor general of Canada, wrote me that they were planning to pay me a visit on their return from Bermuda. "They" meant the governor, Lady Grey, their daughter, Lady Sybil, and Mary Parker. How could I, a poor forester, find the three or four rooms necessary for such illustrious guests in my humble cottage on the Biltmore Estate? In the Pinkbeds, where my shack was extendible like an accordion and where there were camping facilities aplenty, the problem might have been easy and Mrs. Schenck's resourcefulness would have made the Canadians comfortable. Lordy! I had invited the Greys to visit me, but I never thought that they would accept that invitation, four members strong! In that predicament, I asked Mrs. Vanderbilt for advice and help, with the result that we two concocted a letter asking the Greys to lodge during their stay at the Vanderbilt mansion. And they came — charming, modest, and unpretentious.

Earl Grey and I took a number of horseback rides over the estate, killed a few rabbits together, and made a trip to

Pisgah Forest, where his grace, joyous as a boy, caught a few trout in Bradley Creek, getting totally soaked because he insisted on wading in the creek instead of fishing from the bank as we others were wont to do. The Vanderbilts did not have a good car and I had to ask one of my new acquaintances, John D. Archbold, who was summering at the time in Asheville, to convey the party in his fine Packard. A few days before, he and I had taken the first automobile trip ever made through the entire breadth of Pisgah Forest — a trip which proved to me that my rough mountain roads were fit for heavy automobiles, provided the owner did not mind the bumps of the thank-you-ma'ams met at short intervals.

Less enjoyable for me than the Greys were a number of United States senators and congressmen who visited the Biltmore Estate when a Congressional party passed through Asheville.[1] In forestry they showed little interest. They seemed to me weary of sight-seeing before they began. It is possible, too, that they expected to see much more than the Biltmore Estate could show them, having based their expectations on the exaggerated accounts appearing in the daily press. If they were disappointed in the estate, I was disappointed in them. I had expected more interest in forestry than I actually got from them and into them.

The leader of the delegation was Congressman Charles F. Scott of Kansas, at the time chairman of the House committee on agriculture. He surprised me by saying that an Appalachian park was of little use for water conservation, and that it would be better for American forestry to afforest the vast areas of idle and weather-exposed farm land in the Appalachian regions rather than to create parks or forests. In his opinion, however, afforestation was not a national task but a state affair. He added that terrible inundations had occurred in the United States before the advent of the white man, and that no human effort would ever cause them to cease.

[1] The forests committee of the National Conservation Commission.

The most remarkable member of the Congressional party was Senator Reed Smoot of Utah. He was known to be a Mormon, and in my mind a Mormon was a bigamist, a man wearing his hair like Samson and his caftan like King Zedekiah of Jerusalem. And here was a Mormon, by far the best-looking and best-dressed member of the group. I was aghast when he told me that he was well acquainted with my old friends Forstmeister Ulrich Meister in Zurich and Professor Heinrich Mayr in Munich. But I was perfectly nonplused when he said that his daughter was the wife of Count Konigsmark, captain in a regiment of Prussian cavalry. In his questions and remarks, Senator Smoot was the most interested and the most interesting man of the party. "Why, I do not see, Dr. Schenck," he said, "that there is any deforestation at the headwaters of the streams along the mountains framing the horizon. How can you claim that any such deforestation does influence the waters in the rivers and in the valleys?"

On the banks of the French Broad River I could show him a mark made by the Indians on a tree denoting the highest water gauge the river had ever had. The mark was made in 1875, long before the beginning of lumbering in the Land of the Sky. In other words, the highest flood that had ever occurred in the French Broad River had taken place before there was any lumbering at its headwaters. Admitting this fact, I did not mean for a moment to minimize the influence of forestry on the runoff. I meant, and I shall always mean, to minimize any exaggerated statements about it. A good cause may be hurt more by its advocates than by its adversaries.

The most interesting experience in forestry for me during the summer of 1908 was a visit to the Upper Peninsula of Michigan, made upon the invitation of William G. Mather, president of the Cleveland Cliffs Iron Company. This company controlled at the time not only the iron mines and the smelters, the railroads and the harbors on the peninsula, but also its forests, which supplied the mine props, the wood, and the charcoal required for the smelters. In addition, there were sawmills, tanneries, and factories for wood alcohol and for wood vinegar, all working hand in hand toward the destruction of the peninsula forests, in which maple, birch, and elm were the prevailing species.

On the cutover lands immigrants from Finland and Sweden were struggling hard to make a living from a poor sandy soil in a climate wrapped in seven months of winter. Oats, rye, hay, and, if the frost did not come too late or too early, potatoes were the crops produced. There was dissatisfaction among the immigrants because they had been led to expect better conditions. Mather was trying hard to improve the living conditions of the thousands of families connected with his many enterprises. These endeavors, it seemed to me, were doomed to fail, for the workers wanted more money than the economic conditions of the time and the site would permit.

Mather, one of the finest gentlemen I ever met — big hearted, benevolent, socially interested, farsighted — had a wonderful way of talking to his workmen; but the undertone of their replies denoted dissatisfaction. My experience in Pisgah Forest was duplicated by my observations in the Upper Peninsula. Try as one might, the workers will never be satisfied with their working conditions and their wages. "Nor I nor any man that but man is, with nothing shall be pleased till he be eased with being nothing." Who said that? Not a capitalist, nor a workman, nor a philosopher, but a king — King Richard of one William Shakespeare.

On my way back to Biltmore I tried hard, going about like a drummer, to sell some of my unsalable lumber in Detroit, Cleveland, and Cincinnati. All efforts were in vain. The depression, it seemed to me, had reached its lowest possible stage. This was a hard experience for me; and I realized that I did not have the suave qualities required of a successful traveling salesman. The majority of the prospective customers that I visited treated me as if I were a vagabond.

CHAPTER 20

The Biltmore Forest Festival

THE BILTMORE FOREST SCHOOL in 1908 lost its assistant chief, Dr. Howe. His presence in Biltmore had made it possible for me to be absent on trips to Europe, or on trips like the one just described to the Upper Peninsula of Michigan. When Howe left Biltmore to work at the University of Toronto with Fernow, whom he later succeeded, his place was taken by Dr. Homer D. House, a capital botanist of much help to me, since my botanical knowledge was exceedingly meager. And he was loyal to me through thick and thin, in spite of the harsh treatment that he often received at my hands. He was ever ready to help me in the school and in the management of the forest department. I am deeply indebted to him, and since his death in 1949 I have been one with the alumni of the school in cherishing his memory.

On House's advice, I consulted Dr. W. A. Murrill of New York City, who had discovered in the New York Botanical Garden the fungus responsible for the chestnut blight, the most devastating plant disease on record. What would happen if this disease were to kill the chestnut trees of Pisgah Forest, which were the greatest assets I had in the woods? Many of them, it seemed to me, were declining in health; and I feared they were stricken by the chestnut blight. Fortunately, at that time Murrill found no trace of the disease in Pisgah Forest. The blight did not spread to North Carolina until some years after my departure from its glorious woods.

Greater than my fear of a tree disease in those days of 1908, however, was my fear of the election of William Jennings Bryan as president to succeed Theodore Roosevelt. I am certain that Bryan's fantastic ideas in economics did not appeal to my acquaintances among the Ashevilleans, who, of course, were Democrats as a matter of principle. The candidate of the Republican party was William Howard Taft. Should Taft be elected in November, I could hope for a recovery of the lumber business, on which the financial maintenance of my department of forestry depended. No wonder that I advocated the election of Taft in all my correspondence which, in those days, was pretty heavy. A circular letter to my customers, however, in which I recommended Taft's election and simultaneously my lumber, had a decidedly negative result. Many answers from my Democratic customers did not mince words, and they fell on me as if I were a thief or a murderer. Political excitement was at a high pitch, but not on a high level.

To my delight, Taft was elected. His election gave me, among other things, hope for the success of a grand "Biltmore Forest Festival" to be held on Thanksgiving Day, 1908, and the week end following it. Since early in the year I had been making preparations for the festival, which, so far as I know, was the first forestry festival of its kind in the United States. It was to celebrate the tenth anniversary of the birth of my Biltmore Forest School and, incidentally, the twentieth anniversary of forestry on the Biltmore Estate, and it was meant to be an advertisement of the Biltmore Forest School. All state governors and all United States senators and congressmen whom I suspected of having an interest in forestry were personally invited, as were leading lumbermen, editors of lumber papers, and federal and state officials connected with forestry. My letter of invitation, dated September 1, 1908, is so characteristic of me in those years and of my conception of the character of Americans that I cannot refrain from giving it here in part:

Dear Sir and Friend: —

Rejoice with us. On the 26th day of November next we shall celebrate the twentieth anniversary of forestry at Biltmore, together with the tenth anniversary of the Biltmore Forest School.

Rejoice with us, and make our hearts glad by your welcome presence at the forestry festival, beginning on Thanksgiving Day, November 26, 1908, and ending on Sunday, November 29, 1908.

You may have heard something of the farms and of the forests found on the Biltmore Estate.

Now we beg of you: *Come and see them for yourself!*

Whatsoever we possess, in forests and in farms, we shall throw open to you for three days; — and it shall be yours!

The members of the Forest School and the employees of the forest department will be placed at your service, acting as your personal guides during your sojourn.

This Is the Program

Thursday, November 26th: Excursion, in carriages, over the Biltmore Estate, leaving Biltmore Village at 9:30 A.M. Inspection of various forest plantations (some 500 acres) replanted in White Pine, Yellow Pine, Ash, Maple, Oak, Chestnut, Hemlock, Poplar and Walnut, made between the years 1889 and 1905. Study of planting operations in actual progress.

Thanksgiving invocation in one of the finest plantations.

Lunch in the woods.

Drive through natural second growth of Yellow Pine (ten to twenty years old) obtained by successive cuttings on several hundred acres of woodlands; through thinnings and through improvement cuttings in course of progress; through compartments treated as hardwood coppice under pine standards; etc.

At each site, the director will explain the situation and the purpose in view; silviculturally as well as financially, and the means adopted to reach the desired ends.

Evening: 8:30: Gala dinner at Battery Park Hotel.

Friday, November 27th: Excursion, in carriages, over the Biltmore Estate, leaving Biltmore village at 9:30 A.M. Visit Biltmore Herbarium, Biltmore Nurseries, Afforestations on Coxehill, Biltmore Dairy, Biltmore Pig Farm, Biltmore Poultry Farm.

Study of seed regeneration of Yellow Poplar in Compartments 102 and 104.

Lunch in the woods.

Trip to plantations of 1899 (Pine, White Pine, Locust, Cherry, Walnut, Sugar Maple, Basswood, Oak) and also tc those of 1907 and 1908.

In the evening at 8:30: Possum hunt on the Biltmore Estate, barbecue and rejoicing.

Saturday, November 28th: Excursion, by train to Pisgah Forest and in carriages to the mountain forests; study of the logging operations and of the second growth obtained in primeval forests logged conservatively in 1895, 1896 and 1897.

Lunch in the lumber camps.

Fishing and shooting contest in the afternoon.

Do not miss this excursion to the mountain forests!

By contrasts we learn.

The contrasts existing between the management of our forests situated *close* to the market, and the management of our forests situated *far* from the market is most striking: *It will strike you.*

Pedestrians may ascend Mt. Pisgah (5,700 feet) at sunset, spending the night at the Hunting Lodge amongst the stars.

Sunday, November 29th: Farewell!

These Are the Inducements

There is not, and there cannot be offered in the United States an opportunity better than that existing at Biltmore for the study of *Practical Forestry* and of its actual results, obtained with an eye single to success by twenty years of uninterrupted activity! Just think of it! You shall see the forests in all stages of their development, from the embryo forests planted in 1908 up to the primeval woods containing trees antedating by their birth the discovery of America; you will be shown the difficulties, the expense accounts and the revenue sheets of the first forestry on American soil, within an area of 120,000 acres, the property of Mr. George W. Vanderbilt.

Ye Statesmen! Come and assimilate that experience in forestry which the people are anxious for you to possess!

Ye Lumbermen! Come and obtain after seeing the workings of practical forestry at Biltmore, some inspiration for forestal attempts on your own holdings!

Ye Engineers! Come and study the results of afforestation, accomplished on 2,500 acres of fields once abandoned by the farmer!

Ye Foresters! Come and learn from the woods rather than from the books the methods by which trees must be raised and tended!

Statesman! Lumberman! Engineer! Forester! Come! and be welcome!

Then followed complete information as to the expenses connected with attendance, including railroad fares from the larger eastern cities, hotel rates at Asheville, carriage bills at Biltmore and Pisgah Forest, etc., and a final paragraph reading as follows:

The days selected for the festival — selected for your convenience — are apt to be the days of the finest fall weather.

Do not don your best! Select a rough, comfortable suit of clothes and a pair of shoes in which you may walk a quarter of a mile without the sensation of "walking on a toothache."

Come as you are, and take us as we are!

Come! Rejoice with us! And join us in giving thanks on Thanksgiving Day, 1908.

There were three days of it, and the good Lord gave me the finest possible fall weather. There were fifty guests, and all of them seemed to be as enthusiastic as I was. Some of the great lumber, wood, iron, and steel companies sent representatives, as did the lumber papers. Among the latter was James E. Defebaugh, editor of the *American Lumberman* and my personal friend who, I believed, had been converted to forestry. There were a few bankers, as well as forest commissioners of Pennsylvania, New York, Massachusetts, Texas, and Ohio, and, to cap the climax, one real live United States senator representing his ninety-odd absent colleagues. Of Gifford Pinchot's United States Forest Service, not one member accepted my invitation! I was much chagrined over this failure. I saw clearly then that Pinchot's and my ways had parted forever. From that time on I had no connection with the Forest Service except through some of its staff members who came to the Biltmore Forest School during their vacations as lecturers.

For the three excursions, I had prepared a guidebook which described the types of forestry that were to be seen. The descriptions were numbered, and the numbers corresponded with numbers of signposts along the road. Stopping at each numbered site, I explained the forestry operations to be seen; and next to the driver on the box of each horse-drawn vehicle sat one of my senior students ready to answer any questions which the occupants of his carriage might have.

I considered my Biltmore Forest Festival a great success. The enthusiasm of the participants was not feigned. For the first time in their lives they had seen real forestry in America; for the first time they had visited a tract of primeval woods not devastated, but actually flourishing, after lumbering. The men acted like the Greek soldiers of Xenophon when for the first time they saw the Aegean Sea. Actually I had arrived at the highest point of my career as an American forester. From that time on there came a steady decline.

Now what of my hope that with Taft's election there would be a resumption of business in general and of my lumber business in particular? When the new tariff was being debated in Congress, the National Lumber Manufacturers' Association held a meeting in Washington at which I was asked to speak on the connection between forestry and a high tariff on imported lumber, particularly from Canada. This is the gist of what I said:

Forestry is a new industry, and a new industry must be protected by a protective tariff till it can stand on its own feet. The Canadians, employing cheap Oriental labor in the West, and in the East having plenty of waterways for transportation of wood and logs from stump to factory, are in a better position to produce than we are in the United States. By a protective tariff, the people go on record as favoring forestry. Forestry requires high stumpage prices; there is no sense in producing a new crop of wheat, corn, or cotton or of trees unless the new crop is sure to obtain good prices. The countries in which forestry is at its best are countries of high tariffs on imported wood. Naturally, the American daily press, which actually rules public opinion in the United States, is opposed to a tariff because wood pulp and sulfite, its raw materials, are largely imported.

The effect of my orating in Washington was nil. And President Taft, soon thereafter, came out with the worst message against forestry ever pronounced by an American president. He said that we ought to have free lumber; and that, by giving our people access to Canadian forests we would reduce the consumption of our own which, in the hands of comparatively few owners, had a value that required the enlargement of our timber resources.

President Taft forgot to say that stumpage prices had risen in the preceding ten years in Germany, in Sweden, in Russia, in India, in short, everywhere. He forgot that a world timber famine, for which there is no remedy but world-wide forestry, was predicted by leading scientists. He forgot that the number of forest owners in the United States had exceeded the million mark. He forgot that nothing can be conserved that is not worth conserving.

CHAPTER 21

Last Days at Biltmore

Now I come to the year 1909, my last year as forester of the Biltmore Estate. All logging in Pisgah Forest and all woodcutting in Biltmore Forest were carried on halfheartedly. There was no market for the products; the winter was mild; fuel wood was not in demand, and not one of my customers seemed to be in need of lumber. Deeper and deeper went my department in the red — in debt with the banks and with me. For two years I had received no salary, and many a payroll was paid out of my own pocket. I had turned over to George Vanderbilt, as agreed, one-half of all revenues coming to me from tuition fees, from expert work in the courts, from the receipts of C. A. Schenck and Company, and from my salary as forester of the Highland Forest. That one-half was far more than was my Vanderbilt salary, which remained unpaid. Of course Vanderbilt knew about this, because my budgets submitted to him year by year on the first of November made the situation perfectly clear.

The winter was filled, more than usual, with dinner parties and dances and social functions in Asheville, to the satisfaction much more of my stomach than of my mind and soul, which came home from them empty and hungry. Exceptions were the Dickens evenings which Mrs. Schenck and I had arranged on Wednesdays with some close friends, when Dickens was read and discussed from alpha to omega.

In March, 1909, Dr. Charles William Eliot, a man famous as president of Harvard University, as a scholar, and as a public leader, came to Biltmore. He was at the time the spiritual president of the United States. Naturally I was eager to meet him and to discuss with him the Harvard School of Forestry at Petersham, Massachusetts, which had recently been established. Dr. Eliot and the Biltmore pastor, Dr. Rodney R. Swope, took a drive with me over the Biltmore Estate in a comfortable landau; farms, nurseries, and forests were inspected; Mrs. Schenck had us for lunch and we were together for some four hours. I cannot say that I got from Dr. Eliot as much of an uplift in my misery as I had hoped for.

Alas, hopes are like trees; most of them die young, few grow to maturity, and very few bear seed to regenerate themselves within the lifetime of a man. Many trees turn out to be cripples, lingering along and always disappointing. Saint Paul was right in preferring love to hope, for love gives satisfaction without fail; hope does not. And yet, if I had the choice today between Caruso singing or Eliot talking, I would choose Eliot. President Eliot was charming. He was a good listener and a great speaker all in one. There was, however, some disagreement between us. He did not like my students who followed his carriage on horseback as a guard of honor, "because they are said to be immoral and given to drinking." In vain was my defense of them, in vain my explanation of our sängerfests.

As we passed Biltmore House, which in my opinion was at that time the finest private building in the United States, he remarked that he did not like it because "it smacked of feudal society." In a talk on education I expressed the opinion that the measure of education in an individual is his usefulness to society. Dr. Eliot snapped: "Do you mean to say that a shoemaker is better educated than a teacher or a preacher?"

I replied that my remark referred to individuals, not to professions. Speaking of books, I ventured to say that the best book on forestry was that of nature, and that the best observer was the best reader, adding that that opinion might hold good for other professions. To this he replied: "Of course, every good book is an outcrop of observations, and a school book is no good unless it is an eye-opener for additional observations." Finally, Dr. Eliot was visibly displeased when I said, in defense of the German army, that it was the greatest educational institution that the world had ever seen. It was in vain that I tried to explain my attitude on the basis of my own personal experience in the German army.

In the evening President Eliot gave a public lecture in the Asheville auditorium on "The School of the Present and the School of the Future." The new school does not teach by books, he said; it favors individuality and it allows "skippers" to pass through the courses at a rapid rate; the new school discards thoroughness on purpose in favor of progress; it has laboratories, stereopticons, graphs, and collections. Eliot spoke with a much subdued voice. The audience was inattentive and noisy, making it hard to listen. There were a few striking and unforgettable remarks, among them: "Freedom creates, not equality, but inequality"; and "The parallel existence of many churches is desirable."

Talking on education, Dr. Eliot might have borrowed some educational witticisms afloat at the time in the Biltmore Forest School: "Damn the school when it acts like a sawmill in which different logs are made into lumber, all evenly edged, all evenly trimmed, and all kiln-dried alike"; "Marie François Voltaire is right: woman is the riddle of the universe, but notably the riddle of the universities admitting co-eds"; "A college is a boardinghouse where knowledge is served in lieu of meals. Alas, the Biltmore Forest School can cook the meals for its boarders but it cannot force the boarders to eat them"; and "It is hard knocks on the universities, here and abroad,

that men like Edison, Burbank, Walt Whitman, the Wright
brothers, Westinghouse, Elbert Hubbard, *et al.*, have graduated merely from the University of Hard Knocks."

O Lo! What blasphemies! Yet, does not modern education
take the water from the fountain of youth in the hills, put it
into iron pipes, convey it on a downgrade and in the dark
to the city and to the village, liberate it of a sudden through
a million faucets for innumerable purposes, and then is astonished when the water disappears in the sewerage of life?

My great adversary at the Biltmore Estate was now C. D.
Beadle, chief of the landscape department and of the Biltmore Nurseries. Beadle had come to Biltmore in the early
days of the Biltmore Estate upon the recommendation of
Frederick Law Olmsted, who appreciated his talents as a
nurseryman. In the course of the years Beadle had risen to
the position of a department head. He was a good manager.
He got along remarkably well with the foreman of his department, and he acquired a unique knowledge of dendrology
by assembling in his nurseries the thousands of plants which,
according to Olmsted's plan, were to find a place in the Biltmore Arboretum. Vanderbilt, being more interested in landscaping than in forestry, came to appreciate him so much that
he was now, in a sense, the general manager of the Biltmore
Estate, but without any influence in the affairs of my forest
department.

When the Vanderbilt funds began to run short, Beadle
made strenuous attempts to commercialize his nursery, advertising for sale what surplus plants he had to offer from his
landscape nurseries. His customers were the owners of Northern estates and, since spring in the North is some four weeks
later than at Biltmore, his plants were in leaf and in sap
when his Northern clientele received them. That was the
reason, I believe, for Beadle's failure to make his nurseries a
commercial success. When I sold my surplus stock of white
pine seedlings by thousand lots, he complained of unfair com-

petition and he persuaded Vanderbilt to forbid any direct sales of my nursery stock when I had any surplus to dispose of. I did not oppose the action because my sales were made only occasionally.

While the Biltmore Forest School, my forestry office, and my household were in the midst of preparations for a spring and summer to be spent in the Pinkbeds, there came, on the afternoon of Saturday, April 24, 1909, the turning point, not only of my career, but of that of the Biltmore Forest School. That is the date on which forestry at Biltmore came to an end.

It happened thus. In a discussion with Beadle, he had the temerity to remark to me smilingly that I had told an untruth about something — I do not now remember what it was — to Vanderbilt. When he refused bluntly to retract his insult, I boxed his ears as best I could. Thereupon his office staff came to his rescue, and I was forcibly ejected from the premises. The result of the affair was a complaint by Beadle against me for assault and battery. On June 3 Walter R. Gudger, a justice of the peace in Asheville, found me guilty and fined me one dollar. May God Almighty be as lenient with me on judgment day as was Magistrate Gudger on June 3, 1909!

On the Sunday succeeding this affair I followed my school and office to the Pinkbeds, on the way inspecting the result of some forest fires which had been set by moonshiners during the preceding week. In the Pinkbeds tanbark peeling was in full swing, in spite of continuous rains; my logging operations were curtailed because the lumber market was dull; and road building was at a standstill because my exchequer was empty. In this predicament I escaped from all worries by taking a horseback trip, accompanied by some forty forestry students. We rode southward and westward in the direction of Georgia, where the Three State Lumber Company controlled some twenty thousand acres of forest land and where logging operations continued in spite of a dull market. The rivers, notably the Chatooga, were high and therefore fit for driving logs. The loggers, wading in the water, were pushing the logs ahead wherever a jam formed, and many a log had to be rescued from high upon the banks where it had been washed by the turbulent waters. I do not believe that our host, the lumber company, was making any money in the enterprise; anyhow, this excursion was instructive for my students and for me.

On my return to the Pinkbeds I learned from the newspapers that my affair with Beadle had been published widely over the country. My friends told me that there were persons intriguing against me to undermine my position on the Biltmore Estate. The Vanderbilts had just returned from abroad and, no doubt, were worked upon by my adversaries.

There were causes enough for intrigues. To end or to mitigate the financial misery of the forestry department I had closed a contract with the Asheville and Chicago Hunting Club, by which the exclusive rights of hunting and fishing in Pisgah Forest were leased to the club for ten years for an annual payment of $10,000. Naturally, I had consulted the very best lawyers, as well as my friends, before I put my name to this contract, by which I would utilize a source of revenue which I had been developing with a will for twelve years. At the same time the contract would provide a chance to sell, for club holdings and for prospective cabins of the club members, the lumber of an inferior grade which it did not pay to ship by rail to the Northern markets. In addition, my farmers in Pisgah Forest would have a chance to sell their milk, eggs, and vegetables locally to visiting club members and to make a few spare pennies acting as guides for them.

To my amazement, Vanderbilt was furious when he learned about this contract. In vain did I say that I had his written permission for contracts of this type; that he had encouraged a similar contract two years before; and that one thousand acres immediately surrounding his hunting lodge had been reserved for his personal use and for that of his guests. Although he never told me so, he apparently had found a prospective purchaser for Pisgah Forest, and he was afraid that this hunting contract might deter his customer from closing the deal. Whatever the case may be, George Vanderbilt, in several interviews with me, was angry and went so far as to call me an idiot and to demand my resignation.

In this predicament, and on the very day of this catastrophe, there came to me from Europe a cable reading as follows: "Uncle Max dead. Last will leaves you an annuity of 2,500 dollars. Signed Fritz." Uncle Max was the Russian uncle of whom I have spoken previously; Fritz, the signer of the cable, was my brother-in-law, who had visited me in Biltmore four weeks earlier. He was as close to me as a brother. The annuity awaiting me relieved me of immediate financial worry in leaving my position at Biltmore and gave me at the same time the stamina required in my final break with my employer. There was comfort in the fact, too, that my friends in Asheville and in Biltmore, who outnumbered the local friends of the Vanderbilts ten to one, were sticking by me. Most loyal of my friends were the students of the Biltmore Forest School, whose attachment seemed to be increased by the predicament in which I found myself.

Naturally, I was unwilling to abandon the Biltmore Forest School when I gave up my position with the Biltmore Estate. The school was self-supporting in that the tuition fees of the students were paying its expenses, while I myself, on the basis of the cable just received, had enough to keep me going. Since there was no cause for my discharge other than Vanderbilt's change of mind, he was legally held under my original contract of employment to surrender to me the benefits of an annuity insurance policy in the Mutual Life Insurance Company of New York. As a consequence, I personally was independent of any revenue of the school and was in a position to use tuition fees solely for the benefit of the school.

But I had no schoolhouse and, what was worse, I was losing the working field of the Biltmore Forest School on the Biltmore Estate during the winter and in Pisgah Forest during the summer.

What could I do? I had many friends among the lumbermen and I knew that they would help me. If that help consisted of permission to use their holdings in North Carolina, in the Adirondacks, in Michigan, and in Oregon for working fields, the Biltmore Forest School would be able to offer to students better opportunities than ever for an acquaintance with the conditions confronting forestry all over the United States. An additional plan evolved in my mind. Object lessons in sustained-yield forestry were not available in the United States, while there were plenty of them in Germany, in Switzerland, and in France. There was nothing to prevent me from taking my students to Europe regularly for two or three months and showing them on the ground what sustained-yield European forestry looked like and toward what goals American forestry might strive. I anticipated, incidentally, that the Biltmore Forest School would be welcome as an adjunct to the Darmstadt Tech, so that there would not be any need for lecture rooms, botanical collections, and geological exhibitions at Darmstadt.

Naturally, I was sorry to abandon my work on the Biltmore Estate, but why should I not continue to have one of the working fields of the Biltmore Forest School in western North Carolina? I was the owner of two small tracts within Pisgah Forest where it was possible for the Biltmore students to find living quarters with the neighboring farmers. Thus it happened that my entire household and my horses, carriages, and dogs were transferred from the Pinkbeds to one of my own places within Vanderbilt's Pisgah Forest and were left in charge of a reliable keeper. Next spring — that was the plan — the Biltmore Forest School was going to spend a few weeks in Pisgah Forest, in spite of George Vanderbilt's expulsion. When all these arrangements were perfected, the school, some forty students strong, sailed to Europe on the Holland-American steamship "Potsdam." I selected this line because it offered the school the greatest comforts at a minimum price, and I was eager to show my students en route to Darmstadt the operations of the most efficient forestry association of the world, those of the Nederlandische Heide Maatchappij in Holland, from where we had the best of rail connections to Darmstadt.

Obviously, when the Biltmore Forest School sailed for Europe in the fall of 1909, my best years were over. I was forty-one years old. What did I leave behind me? What had I accomplished in the United States since my arrival in 1895?

I had established the first school of forestry in America.

I had established on the 7,500 acres of Biltmore Forest what is now known as the first private forestry in America. Biltmore Forest was made fit for a sustained annual yield as reliably as any forest in central Europe. There was nothing like it at the time in the United States.

Pisgah Forest's hundred thousand acres of woodlands had been utilized conservatively, had been supplied with a system of roads, and had increased tremendously in value. There was regular work in it for the natives. The woodlands had been well protected from fire and from trespass.

The backwoods families living in Pisgah Forest had been supplied with schools, with church and postal services, and with a small amount of the comforts of civilization.

I had proved that logging is the leading part or function of American forestry, that forestry and lumbering are inseparable.

In my various books I had shown for the first time in America what scientific and practical forestry look like, and that there are two great legitimate sections of forestry: (1) public forestry, or forestry undertaken by the people in nation, state, and community, as subdivided into branches of forest influence, or role of the forests, forest policy, forest legislation, including criminal law, forest statistics, and forest research; and (2) private forestry, or forestry undertaken by private owners, as subdivided into branches of (a) production of wood goods — logging including road building, utilization including lumbering, silviculture, or the art of producing and tending a forest, and forest protection, and (b) management of the producing factors — mensuration, including scaling, cruising, and grading, forest finance, surveying, administration, and working plans and technology, including inspection rules and lumber grading.

For the first time in history, I, as a forester, had advocated the advisability of destructive forestry in preference to conservative or to constructive forestry wherever there was no chance for the latter to be or to become remunerative.

I had secured for North Carolina some laws favorable to forestry. Upon my request, Governor William W. Kitchin had issued a proclamation in May, 1909, placing all woodlands situated above contour line two thousand feet under increased fire protection. I thought at the time that a new era for forestry in the United States was thus initiated.

In contrast to the policy advocated by the United States Forest Service, I had advocated the necessity of an adequate tariff on imported lumber, forest products, and newsprint. Unfortunately, American democracy had failed to prove by its tariff laws that it was in favor of forestry as an industry.

I had had published — some of them after parting with George Vanderbilt — a number of textbooks based on my experiences with American forestry. They were printed for the benefit of the Biltmore Forest School. Their merits lie in their priority and in nothing else. Most of these books are to be found in the libraries of the United States Forest Service, the Yale School of Forestry, the New York State College of Forestry in Syracuse, and the American Forest History Foundation of the Minnesota Historical Society in St. Paul. In

Experimental Stand of Douglas Fir in Heidelberg, Germany
The white crosses, a yard long, indicate the size of the thirty-five-year-old trees.

Biltmore Special Train near Portland, Oregon, 1911

*Biltmore Students Observing Western
Logging Methods*
On Chapman Timber Company lands in Oregon, 1911

addition to these textbooks, I have on my bookshelf quite a number of bulletins and reports published in those early days of American forestry which might be of interest to the historian.[1]

My influence in the United States might have been greater had I become an American citizen. Why did I not adopt America for my country? These were my reasons:

My mother, who was my idol, was German through and through, and she was anxious for me to remain a German.

I had no children to be educated in the United States.

What I had seen of democracy in Asheville and in North Carolina did not impress me so favorably as to cause a preference for it over that constitutional monarchy under the grand dukes (landgraves) of Hessen, under whom my ancestors had served since 1619. And I admired Kaiser Wilhelm II as a Christian monarch and as the man who had maintained world peace from 1888, when he became emperor, to 1909 — twenty-one years!

I was enamored with Lindenfels, the summer place of my parents, and I wanted to pass the end of my days there.

The very possibility of gaining an advantage financially by a change of nationalities seemed to me to be abhorrent.

I love my native country, Germany, as I love my mother; I love my adopted country, the United States, as I love my wife. And why should a man be forbidden to love mother and wife with an equal ardor? Why is he expected to abandon one love for the other?

As it turned out, my German citizenship stood me in good stead after I left Biltmore in 1909 for a winter term in Germany. I had no difficulty in obtaining the position of an assistant professor at Darmstadt Tech, thereby associating my Biltmore Forest School with that well-known institution.

[1] A selected bibliography of Dr. Schenck's published books and articles may be found on page 211, below.

CHAPTER 22

The Odyssey of
the Biltmore Forest School

THE STEAMER CONVEYING the Biltmore Forest School to Europe in the fall of 1909 was comfortable, but slow in a rough sea. The students attended my daily lectures in the dining room of the second cabin as regularly as if we had been on land. There was lots of time for studies and lectures on the high sea, and I myself had more time to prepare my lectures than I had ever had on land. Lordy, if I but had the funds — so went my thought — I should love to have a steamer for a forest school meetinghouse.

And now having embarked the reader on a grand new adventure in American forestry, let me describe briefly in retrospect the odyssey of my beloved Biltmore Forest School in the years from 1909 to 1914. The school had six working fields: Germany, with headquarters in Darmstadt; France, with headquarters at Mimizan-les-Bains, Landes; New York State, with headquarters at Tupper Lake; North Carolina, with headquarters at New Bern in the east and Asheville in the west; Michigan, with headquarters at Cadillac; and Oregon, with headquarters at Marshfield, now Coos Bay. Moving from one working field to the other, the students inspected all sorts of forests and wood operations en route. Every working field had room in itself for continual excursions and explorations. In addition, there was opportunity at every working field in the United States for the students to establish personal contacts and friendships of importance to them in later life.

GERMANY

In Holland, close to the German frontier, was Arnhem, seat of the Nederlandische Heide Maatchappij, a forestry association unique in that it served its members by doing their afforestations and their forest thinnings and conducting their timber sales. In addition, it was a training school for members who desired short cuts of instruction in the kinds of forestry adaptable to Holland. From Arnhem the Biltmore Forest School proceeded up the Rhine and passed by Darmstadt to stay two weeks in Lindenfels, the summer place of my family. Here I was a good guide because for forty years I had known the surrounding woods, some owned by the village, some by the state, and some by the farmers. There were splendid lessons in silviculture to be studied — stands of beech and oak of all age classes up to two hundred years. In Lindenfels, as in the Black Forest, the first forest industry was glassmaking, for which charcoal was needed as a raw material. The oldest plantations of conifers, all of which were newcomers to the Odenwald, dated back to 1804.

While we were busy in Lindenfels my German assistant, Richard Kern, made arrangements in Darmstadt for housing the students and for furnishing our "schoolhouse," formerly the residence of the Mercks of New York City, which the family had generously placed at our disposal. Situated as it was in a dendrological park, with a palm house containing two carloads of American lumber to be inspected by the students on rainy afternoons, the Merck house was a really ideal schoolhouse. The forests, surrounding Darmstadt as if it were an egg in a nest, were in its immediate proximity.

And there was the Darmstadt Institute of Technology, which

ranked high, notably for its courses in physics, aeronautics, papermaking, engineering, and architecture. The Biltmore Forest School became affiliated with the institute through the botanical division, headed by Dr. Heinrich Schenck, a close friend but not a relative of mine. Dr. Schenck was famous internationally as one of the coeditors of Strassburger's *University Manual of Botany*. Connected with the botanical institute of the Tech is an arboretum, now about a hundred years old, containing many American and Japanese rarities.

To the west of Darmstadt stretched the sandy plains of the Rhine, stocked with endless forests of Scotch pine. When the old stands were mature, at about a hundred and twenty years, they were clear-cut in small strips and replanted at once with seedlings from one to two years old. At the time of our visits, Irish potatoes were grown for a year or two between the rows of planted pine seedlings. Incidentally, under this system of silviculture combined with agriculture — known in the teak region of Burma as toungya — the pine seedlings are well protected against weeds, ground fires, and diseases. All pole stands of pine were underplanted with beech (nowadays with American red oak) to improve the humus layer and to shade the soil. The Germans, instead of eradicating the oaks and other hardwoods in the pineries as is done in America, are spending time and money on their increase.

To the east and south of Darmstadt, in a hill country and on a more loamy soil, the hardwoods, notably beech and oak, prevail. The German beech is almost free from defects and is an ideal tree for veneers and railroad ties. When the present mature stands, which are now from 120 to 150 years old, were babies, no one foresaw the advent of railroads needing ties and of furniture works requiring veneers. In the absence of coal, what the people wanted was fuel wood, and beech was grown far and wide to supply the present generations of man with it.

From Darmstadt as a center, day excursions were made to the city forests of Heidelberg, an hour's ride by rail. With the history of the forests recorded in detail for some four hundred years, and with maps illustrating their gradual development, rising out of sheer wastes, the Heidelberg forests were particularly instructive. In these forests today are large stands of American white pine, doing particularly well in sheer rock gardens, and of Sitka spruce, red oak, and hemlock; and there is a pole stand of Douglas fir now some eighty years old, unpruned, but almost free from side limbs. It might be worth while to propagate this Heidelberg strain of Douglas fir in the United States. At the entrance to Heidelberg Castle are two American black walnut trees over two hundred years old, better than the best that I have ever seen in the United States.

Here I should mention some timber groves in Germany which were then and still are of special interest for Americans, be they foresters or laymen. Count Berckheim, a German aristocrat who owned a small estate in Weinheim halfway between Darmstadt and Heidelberg, had visited California some forty years before. Instead of gold he found trees — sequoias, Lawson cypress, incense cedar, red cedar, white fir, Douglas fir, sugar pine, ponderosa, and Jeffrey pine. So impressed was he by their grandeur that he brought their seeds back to his German nurseries and planted pot-raised seedlings with the utmost care on a few dozen acres of his lands between 1870 and 1875. The result is the most fascinating. forest arboretum in the world. The Biltmore Forest School was a frequent visitor to these plantations, distant from Darmstadt no more than thirty miles. No American visiting Germany should miss a visit to this arboretum.

Close to Darmstadt the students had access to the town forests of Frankfurt, with the oldest planted stands of Scotch pine in the world and some planted stands of white pine a hundred and twenty years old. In the Ysenburg forests near by we saw stands of Scotch pine raised from self-sown seeds — then a curiosity in Germany although prevailing everywhere else in the world. What is the explanation? Scotch pine was a newcomer in the west and south of Germany. It could not be raised from mother trees in that section of Europe because there were no mother trees. Scotch pine was planted universally in lieu of beech, which the foresters would have preferred, but beech refused to grow on its old homeland when the soil became impoverished by centuries of misuse.

The Biltmore Forest School spent two or three winter weeks regularly in the heart of the Black Forest. There the Kurhaus Sand served as headquarters. It supplied also the schoolroom for the lectures unfailingly delivered in the forenoons. The forests surrounding us there were state forests of the Grand Duchy of Baden, with Norway spruce and silver fir as the leading species. Splendid systems of roads had taken the place of creeks and rivers formerly used for splashing logs and driving wood. There were many stands of prime timber averaging a hundred thousand feet an acre. All regeneration of the stands, removed gradually over a period of some thirty years, is by self-sown seeds falling from the mother trees. This process of silviculture is badly imperiled by hurricanes, which play havoc in the stands of the mother trees as soon as they are partially opened.

Most hurricanes in the Black Forest come from the southwest. As a consequence the wise forester begins his cutting at the northeastern edge of a stand, allowing his operations to proceed in the direction against the endangering storms. This practice had been developed systematically in the Württemberg part of the Black Forest, in keeping with teachings — and with the commands, when he was chief of forestry — of Christoph Wagner, author of a book on "the geometric organization of the woods," which was much admired, but not by me. The Wagner system, its merits and its demerits, were seen by the Biltmore Forest School in the state forests of Schönmünzach and in the family forests of the Counts of Limburg at Gaildorf, where the Wagner innovations had originated.

If I were writing a treatise on German silviculture, it would be incumbent on me to describe the protective "wedge systems" of silviculture connected with the names of Dr. Eberhard in Langenbrand and Karl Philipps in Huchenfeld, in districts situated at the northern edge of the Black Forest. It is sufficient merely to say that all these exhibitions were visited by the students of the Biltmore Forest School. If they did not assimilate anything from this kaleidoscope of silvicultural types, they certainly learned the lesson that the practice of silviculture is based on factors absolutely controlled by locality and by time. There is no such thing as "the best sort of forestry" regardless of time and site.

For me personally these trips took me back to the forests that I had first seen almost twenty years before as assistant to Sir Dietrich Brandis, when that grand man was piloting scores of English-Indian foresters through the woods of Central Europe. Our excursions, however, differed from those arranged by Sir Dietrich in that we did not travel about continually as if we were gypsies. Instead, after establishing ourselves in comfortable headquarters, we radiated in all directions, getting back to headquarters every evening. There were lectures in the forenoons for at least two hours, and on rainy days, for several hours. And my American students had time to themselves for pleasures and sight-seeing.

SWITZERLAND

From Germany the school's next working field was France, although we did not visit that country as regularly as some other fields. En route to France we traveled through Switzerland. For European forestry Switzerland was and is trumps. There, by its forestry, democracy has shown its superiority over all other forms of government. At Thun we saw the best examples in the world of selection forests owned by small communities, and in Zurich we saw the Sihlwald town forest, with its intricate system of transportation and with the best fuel wood plant that I have ever seen. And there was Couvet, near Neufchâtel, where M. Biollet had abandoned all theories of rotations and age classes and thinnings from above, relying, as if he were a good merchant, merely on his timber inventories carefully kept as his *méthode du contrôle*. Swiss forestry is unique in that the Swiss Federation assists the private owner financially when he constructs permanent lines of timber transportation. There cannot be any sustained forestry without them. We might follow the Swiss example in the United States on the 140,000,000 acres of small woodlands owned by farmers. Gifford Pinchot may have had this course in mind when he said that it is better to help a poor man make a living than to help a rich man get richer.

FRANCE

The forests we saw near Paris, notably at Versailles, St. Cloud, and Fontainebleau, were parks maintained for aesthetic purposes. The woodlands of Chantilly were an exception. There, in some five thousand acres of hardwoods consisting largely of coppice sprouts, we had an opportunity to become acquainted with the "loggers." I recall one trip when a married couple working on Sunday invited us to inspect their home, a shack built of branches and moss. The woman of the one-room house drew a curtain away and exclaimed cheerfully: "Voici les caves de la famille!" A small drink of wine was kindly offered. And this hospitality within eyeshot of the Eiffel Tower and within an hour's ride from the temptations and luxuries of Paris! Indeed, these loggers, like the modest folk of Mimizan, were really French.

It was not Paris, however, but Mimizan-les-Bains that was the mecca of our pilgrimages in France. Situated halfway between Bordeaux in the north and Biarritz in the south, on the Bay of Biscay in the midst of the pineries, this hamlet for decades had supplied Europe with naval stores such as turpentine and resin. The "Landes of the Gascogne" are a poor, perhaps the poorest, section of France. The soil is sand, almost sterile, and utterly unfit for the production of crops. We were told that this country was barren when Napoleon I came to power and insisted on afforesting the wastes with what is known botanically as *"Pinus maritima."* Mimizan itself is one of the tiny watering places frequented by the French families of *épiciers* too poor for the luxuries of great sea resorts such as Biarritz or Boulogne.

What did we want to see and learn at Mimizan-les-Bains? Were there not pineries enough in our own South? And were not the North Carolinians in particular known as "tar heels?" On my tours with Sir Dietrich Brandis I had never seen the French methods of turpentining, and I wanted my Biltmore boys to see them and to learn something from the long experience of America's competitors in the naval stores industry. The French for years had been using the cup-and-gutter method of turpentining, which at the time of our visits was beginning to replace the old-fashioned boxing methods in America. In their young stands of pine, the French knocked off with the ax the green lower limbs of the best trees, and when the trees were about four inches thick the French *raisinier* began to work on them, one small "face" at a time, until the trees, irregularly swollen, were about a foot thick and about eighty years old. Small portable band-saw mills were then used to convert them into railroad ties and boards.

UNITED STATES

The Adirondacks

From the Landes our traveling schedule took us back to America for a swing through the United States, our starting field being in the state of New York. Here our annual headquarters in early spring, immediately after our arrival from Europe, were in the heart of the Adirondacks, close to the woodlands at Axton where Dr. Fernow's forestry venture had met with disaster. Trips to these areas, to the Axton nurseries, to Fernow's plantations, and to his regeneration cuttings were highly instructive, as were the logging and driving operations of the Santa Clara Lumber Company, the Norwood Manufacturing Company, and the Brooklyn Cooperage Company. At Saranac Lake and Lake Clear the students visited the state fish hatchery and the large state forest nursery under Clifton R. Pettis — the finest forest nursery then existing in the United States. In the old woods there were stumps of white pine and spruce in decrepit stands of beech; and there were the Fernow cuttings, many of them blackened by forest fires. Living in the Adirondacks was cheap. The students had rooms and three good meals at the Iroquois Hotel in Tupper Lake for $1.25 a day and the use of the schoolhouse there for daily lectures.

On our way south we stopped in Washington, D.C., for several days, including a Sunday during which the students had time to see their capital city and to visit some of the federal forestry leaders, including Henry S. Graves, chief forester, William B. Greeley, assistant forester, Albert F. Potter, and William L. Hall, who gave us short talks on their particular work.

Biltmore Revisited

On the morning after leaving Washington, the Biltmore Forest School arrived in New Bern, North Carolina, where it was welcomed by C. I. Millard of the John L. Roper Lumber Company. On the Roper holdings at Camp Perry we had a glorious chance for contact with logging and lumbering in the pineries and with the naval stores industry of the section. From there the Biltmore boys had opportunities to cruise and study some cypress swamps along the New River. What is most unforgettable to me about these swamp cruises is the pluck and undaunted courage with which my boys cruised a swamp as readily as they did a stand of longleaf pine.

Numbers of hours were spent in the large mills of the Roper Lumber Company at New Bern, North Carolina, where for the first time we saw lumber dipped in baking soda in an attempt to prevent blue stain; the box factories of the American Can Company; and basket factories using black gum and red gum as raw materials. In the harbor of Norfolk we saw the export shipping facilities for logs (yellow poplar) and lumber.

Our next annual stop, from 1910 on, was in the proximity of Biltmore and Asheville. To my amazement, George Vanderbilt had taken possession by force of my places, the Cagle and King farms, within Pisgah Forest. My furnishings, my horses, and Mrs. Schenck's cow had been removed. All my preparations for the return of the school had been destroyed. Here indeed was a serious situation. Where was I to go with the Biltmore Forest School?

In this predicament my old friend Reuben B. Robertson, manager of the Champion Fibre Company at Canton, North Carolina, some twenty miles from Asheville, invited me to take the school to the brand-new village of Sunburst, which the company was building, but not using pending the beginning of logging operations in their forests. The village was located on the Pigeon River, with Mount Pisgah immediately east of it. My misfortune had resulted in a gain. There were in Sunburst plenty of rooms for the students, four brand-new cabins for the teaching staff, and the most modern schoolhouse in western North Carolina, all made available to the Biltmore Forest School by Robertson's magnanimity.

In Sunburst we had access to the very best of working fields — hardwood and spruce forests, logging operations in the making, sawmills under construction, splash dams in operation — everything needed for the eyes of an American forest school. In addition there was a most up-to-date pulp mill where, as I have previously related, tannin was extracted from chestnut wood and the wood was then cooked by the soda process. Some of the pulp was used in the manufacture of United States postal cards. The extracted tannin went to American tanneries and, via Hamburg, to those of Europe. And in the woods a forester was employed to guarantee the perpetuation of the wood supply. Obviously, the Champion Fibre Company had changed its attitude since that memorable visit of its founder to the Pinkbeds.

Actually we were better off in Sunburst than in Pisgah Forest, with this difference: in Pisgah Forest I had been boss, directing everything with the help of the students; in Sunburst we were guests and mere observers.

There was no welcome for us at the Biltmore Estate, since Vanderbilt had declared war on me. Naturally I was anxious to show the students my old and new plantations, my seed regenerations of pine, my compartments of oak coppice under pine standards, and other forestry work I had been carrying on there for fifteen years. What did I do? Like a thief, accompanied by the students, I climbed the fences, and we roamed all over my old woods, unmolested and unseen, although I myself was troubled with a bad conscience. On Bear Creek an acre of my plantation had been burned, but the German larches had done particularly well. In the Lone Chimney plantation sugar maple was luxurious. In the Old Apiary plantation the white pines that had been planted there close together were free from limbs. Douglas fir and Sitka spruce had done much better than I had expected. By-planting of pine had been of no help to my plantations of black walnut. Lordy, what pains I had spent on those plantations, on every compartment of the Biltmore woods! What had become of the records of every compartment which I had carefully kept for those fifteen years? To this day I do not know.

On the advice of my lawyers, my lawsuit with George Vanderbilt was compromised. I signed a deed conveying to him all right, title, and interest which I had in any interior holding of Pisgah Forest; and he paid the salary due me for 1908 and 1909 and surrendered to me, in accordance with my original contract, a life insurance policy and an annuity policy in the New York Life and the Mutual Life Insurance companies. More magnanimous than Vanderbilt were the owners of Highland Forest in Haywood County, which I had administered for the preceding six years and which had gained so much in value that it could be sold at a price which I thought was fantastically high. Unasked, I was paid a premium of several thousand dollars. On leaving western North Carolina for Michigan in early August, I was thus in a position financially to maintain the Biltmore Forest School and to continue my work as a teacher.

Cadillac, Michigan

On our way north we visited some lumber and veneer mills in Cincinnati, the plant of the Champion Coated Paper Company, mother of the Champion Fibre Company. Near Fort Wayne, we inspected the catalpa plantations developed by the Pennsylvania Railroad. In Grand Rapids the Biltmore Forest School was entertained at lunch at the Kent Country Club by T. Stewart White, father of Stewart Edward White, whom we knew and loved for his logging novel, *The Blazed Trail*. We saw a number of the furniture works which have made Grand Rapids better known in the world than all the white pine logs that ever came down the Grand River.

A night ride in tourist cars took us to Cadillac, Michigan, where we were the guests of the Cummer-Diggins Lumber Company, at that time leaders in the northern hardwood lumber industry. William L. Saunders, general manager of the company, was at the time president of the Michigan Hardwood Manufacturers' Association. In Selma Township, eighteen miles away, a tent camp with a couple of fine cooks and a boxcar fitted to serve as my office awaited us. A mile away from the camp, in some abandoned fields, was an old country schoolhouse available as a lecture room; and all around us were primeval woods of sugar maple, beech, elm, ash, basswood, hornbeam, and hemlock. The finest trees in them were American elms, some of them five feet through, as impressive as the best yellow poplars in Pisgah Forest.

Everything was new for us — the trees, the logging operations conducted by narrow-gauge railroads and by high-wheelers, the life in tents, the food in a logging camp. The whole school was extremely happy over all this newness. For lectures in the forenoon we had Professor H. O. Allison on animal husbandry, Royal S. Kellogg on forest utilization, Dr. Homer D. House on dendrology, Dr. Hermann von Schrenk on mycology and timber treatment, and Dr. Schenck on everything else.

The reader may ask what advantage, if any, did our various hosts obtain from the presence in their woods of the Biltmore Forest School, be it in Camp Perry, at Sunburst, or near Cadillac. I can only surmise. Obviously the hosts had a good opinion of us and they did not want the Biltmore Forest School to die an untimely death when it was expelled from Biltmore by George Vanderbilt. Moreover, they could not help getting through us a good deal of advertising in the lumber papers, free of charge; and who is the lumberman who can get along without advertising when the markets are dull? Our kind hosts also may have had a vision in those early days of the coming possibilities of forestry on their lands; indeed, the perpetuation of their operations by one scheme or another was frequently discussed with them. What work could we do for our hosts? Not much. We made some cruises for them, ascertaining the amount of stumpage available in this or that tract in which they were interested, and they were amazed when they saw the accuracy of our Biltmore methods of cruising.

Among our paid visitors in 1912 was a man nationally known. I wanted my boys to come in contact with Ernest Thompson Seton, whose books on the wild life of the north woods were best sellers. Fred Diggins was afraid that the Biltmore students would not get "my" money's worth out of this celebrity. I had to pay him a thousand dollars for his presence for one week. Seton, who appeared in the dress of a medicine man, demonstrated the Indian trick of starting a fire by rubbing together two sticks of dry wood; he told the students some of his new stories on "animal heroes"; he lectured on the Indian sign language, on the ten commandments in animal life, and on the message of the red man, which is: "Return to Nature!" An unforgettable lecture was one on Buddhism and reincarnation, in which Seton claimed that "will power is God's power," that "will for the good is God," and that "a mind concentrated in prayer works wonders." From his one week with us I personally got a thousand dollars' worth of wisdom out of this remarkable philosopher.

The most interesting experience that we had in the several summers spent near Cadillac on the Cummer-Diggins holdings occurred on the morning of August 25, 1910. I was lecturing at the little country schoolhouse. Suddenly there was a great hullabaloo; the wind howled, rain began to drive through the roof, and the building was lifted from its foundations. Two students were granted permission to leave in order to fasten their tents, which they thought were flying away in the storm. They could not open the door of the classroom, for the storm had distorted it. They left through the window, and five minutes later returned terror-stricken with the news that the entire section of the woodlands where we were tenting was being blown over by a hurricane! Lordy! In their tents in the woods were my brave wife and Mrs. Kern, my assistant's wife.

The students and I lost no time going to their rescue. We had to escape from the building through the window. We followed the wide track of the logging railroad, and I soon discovered that many trees had been blown down. Mrs. Schenck and Mrs. Kern, who had wisely taken refuge at the railroad siding under a car heavily loaded with logs, were unharmed. Our tents had been blown down, and one of them was buried under a big maple tree. No lives had been lost, Our eating shack and the cooks in it, too, were intact.

This storm, we were told, was unprecedented in the annals of Cadillac logging. But, looking over the woods more searchingly than I had before this adventure, I found that everywhere there were small hillocks overgrown with trees, indicating spots where an earlier generation of trees had been storm-thrown. Yes, catastrophes belong to the curriculum of the woods and their trees. Those glorious elms, dotting the stands of beech and maple, had developed, no doubt, in gashes and gaps made by storms some four hundred years before. And there was another demonstration in the woods of the effect of catastrophes. Where a surface fire had bared the soil in winter there were nurseries of young elms several acres in size. Now, after the windfalls caused by the hurricane, some of these elm seedlings would get a chance to grow. Who is the forester capable of solving the millions of puzzles that he finds in every neck of every forest, old or young, intact or partially destroyed?

Before leaving Cadillac the first year of the Biltmore odyssey, I made an agreement with the Cummer-Diggins Lumber Company by which it would supply us with a new bunkhouse and a new schoolhouse on Lake Dayhuff within its holdings, the Biltmore Forest School footing one-half of the expenses involved. It was at this new site that the Biltmore Forest School spent the three subsequent summers most comfortably.

The Pacific Northwest

The transcontinental railroads, so it seemed, were interested in the Biltmore Forest School. Their travel agents offered all sorts of advantages if we would select their particular routes. On every trip west we traveled in comfort, with two Pullmans and a dining car for our private use.

Biltmore Forest School Clubhouse

Sunburst, North Carolina

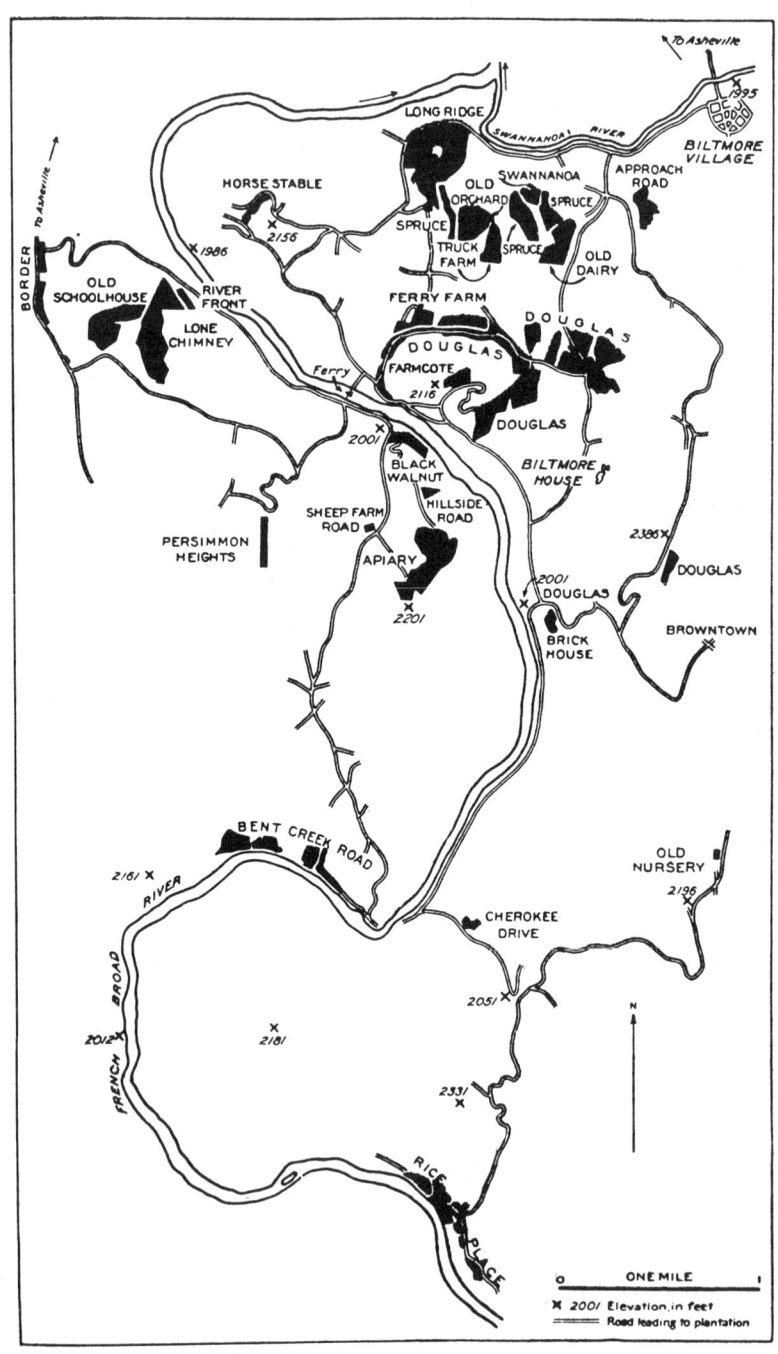

Forest Plantations, Biltmore

[From U.S. Department of Agriculture, *Miscellaneous Publication* No. 61
(Washington, 1930).]

En route to the West Coast we stopped in Duluth to visit the Clyde Iron Works, manufacturers of the McGiffert skidder and, more recently, of a four-line skidder, and Duluth harbor, where the ore ships in the docks were being loaded automatically. We also visited the Chippewa National Forest, near Cass Lake, where fifty per cent of the trees left for seed had been blown down. Management of the national forest was handicapped by the intermixture of state swamp lands and Indian allotments. Our next stop was at Glacier National Park, where we spent the night at the Swiss Chalet, with meals at fifty cents served by Munich waitresses. The next day we were shown over the park. Never had I seen any woodlands of greater density — western larches a hundred and twenty and more feet tall, towering over western red cedar, western white pine, western hemlock, mountain white fir, and Douglas fir. In the coves were gigantic solitary cottonwoods and birches a foot thick. What lectures in forestry these stands were preaching!

The next morning we awoke in Spokane. What a change! All slopes were covered with ponderosa pine. We went on to the Big Bend country of the Columbia River, following the north bank road — a sagebrush country with some irrigated farms, many sand dunes of loess, and outstanding palisades of basalt — and then to the Narrows and to the first Western forests. Is there any train ride in the world more beautiful than was ours?

Arriving in Portland, we were welcomed by George H. Cecil, a Biltmore graduate of 1904 who was in charge of the national forests of the Pacific Northwest; George M. Cornwall, editor of the *Timberman*; F. A. Elliott, state forester of Oregon; and many others. Here began our tours of the wonders of logging and lumbering in the Douglas fir woods in the vicinity of Portland. One of our hosts was the Chapman Timber Company, owners of some eighty thousand acres of fine forests where three hundred thousand board feet of logs a day were being cut and delivered by rail to the river. We visited the mills in the Willamette River Valley, among them those of the Eastern and Western Lumber Company; the Willamette Iron and Steel Works, manufacturers of the best logging machinery of the time; the Oregon wood distillation plant; and James D. Lacey, foremost timber cruising firm of the world, appraisers of timber sales involving millions of dollars.

I cannot describe all the sights in and near Portland that we inspected, such as the St. Helens creosoting plant, Crown Columbia Paper Company at Oregon City, the nurseries of the latter at Clackamas, and their poplar plantation on the river, which produced, we were told, as many as six cords of pulpwood per acre a year. I wish to say, however, that in other years of its migratory life the Biltmore Forest School, going west by the Northern Pacific, had equally wonderful experiences in the Coeur d'Alene National Forest in Idaho, in the forests near Seattle, in the Olympics, and in Mount Rainier National Park.

We went regularly by steamer from Portland to our next working field, Marshfield, Oregon, the harbor city of Coos Bay. Our host in Marshfield was a famous lumber company headed by a visionary, Charles A. Smith, who had been successful as a lumberman in Minnesota. His chief helpmate was Arno Mereen, vice-president of the company and builder of the then largest band-saw mill in the world, at the point where the Coos River enters Coos Bay. More important for the Biltmore Forest School than the head men of the C. A. Smith Lumber Company were two others of the firm: Cornell Lagerstroem, Smith's nephew, and John Lafon, both graduates of the Biltmore Forest School. Lafon was one of my star graduates, and under my general supervision he had earned his spurs as forester for the Highland Forest Company in North Carolina. His help during the six weeks that the Biltmore Forest School spent annually on Coos Bay was invaluable. He secured a classroom for us, first in the "tabernacle," and then in the new and large public school building of Marshfield. I had an office room and a stenographer in the Cook Building, and Mrs. Schenck was most comfortable in a furnished house supplied by Lafon. The Biltmore boys had no difficulty securing rooms and board in a city as hospitable as was Marshfield.

The first thing we undertook in the woods was a trip to the logging operations of the Smith Lumber Company up Coos River, where we saw the construction of railroad extensions; scraper work and pile-driving; huge logs being pulled over a mountain; three miles of trestles, requiring three hundred thousand board feet of lumber to build at an outlay of seven thousand dollars a mile; donkey engines unloading seven thousand feet of logs into the river in exactly ten minutes; and so on. At the city waterworks we cruised a stand of fine second-growth Douglas fir and found that it had a very high average of board feet to the acre. In the depressions there was no Douglas fir, only hemlock and spruce. How to explain this phenomenon? Was it that Douglas firs could not successfully seed the depressions because the matting of duff and humus under them had not been destroyed sufficiently by fires to expose the bare soil? Or was it that the seedlings of Douglas fir were outgrown in the vales by weeds and brush, the shade of which the spruce and hemlock could endure for many years and succeed in the end?

In Marshfield as early as 1911 we were shown the first electric tree-felling saw ever constructed, a small motor driving a chain equipped with sharp teeth, exactly after the principle now employed in the felling saws all over America. The inventor was John Mereen, son of Arno Mereen. He was ahead of his time also in inventing an electric donkey, which was constructed by the Westinghouse Electric Company. Verily, there was not a day during our stay on Coos Bay when my students and I myself did not learn something new and instructive — not a day when I was not presented with problems for which I had no explanation.

Of course our excursions on the shores of the Pacific Ocean were not restricted to the holdings and enterprises of C. A. Smith. We saw the veneer mills, the shipyards, and the sash and door works at North Bend, and the old sawmill of the Simpson Logging Company, using two circular saws, each five feet in diameter, for sawing the largest logs of Douglas fir and cedar. On Daniels Creek we saw the operations of McDonald and Vaughan, who used standard-gauge cars on an incline a mile and a half long for getting out the best logs that we had seen. What struck us most, however, was the fact that C. A. Smith was ahead, not only of the others, but also of his time. Furthest ahead of its time and ruinous for Smith's companies was a sulfate fiber plant, built of concrete, in which, after Norwegian patents, two Norwegian youngsters promised to produce from the waste of the sawmills a high grade of pulp. The plant, with digester walls one and one-eighth inches thick made of welded steel and costing hundreds of thousands of dollars, never turned a wheel. Smith, a Norwegian by birth, was badly misled, I am afraid, by his own countrymen.

After leaving Marshfield, on the way back to the East the Biltmore Forest School visited the logging operations of the Polsen Logging Company at Grays Harbor, south of the Olympics, where Frank H. Lamb, an early Biltmore graduate, had made a first attempt at logging conservatively. He left large stands of second-growth Douglas fir and spruce, which, unfortunately, were destroyed by fires a year later. At Tacoma the school was the guest of the St. Paul and Tacoma Lumber Company, headed by Major Everett G. Griggs, president of the National Lumber Manufacturers' Association. There, at Kapowsin, we saw the Lidgerwood aerial skidder — the new "flying machine" of the loggers — the gigantic chutes conveying equally gigantic logs, the new roaders, kick-overs, and yarders — everything supermodern and superefficient. There was another startling innovation. A. F. Wilbur, the logging superintendent, was working in two acts. In act one, the under story of small trees, such as cedar poles, was salvaged with old-fashioned horse teams before the giant Douglas firs of the upper story were cut in act two, smashing whatever under story there might be left.

In Seattle we saw the forestry school of the University of Washington, which had been established in 1907; the giant establishments of the Pacific Creosoting Company; the Western Crossarms Company; the Seattle Cedar Manufacturing Company; and the Pacific Coast Pipe Company, where huge wooden pipes were being made of staves jointed by double triangular tongues and grooves. And, wonder of wonders, the woods operations of the Merrill and Ring Logging Company, employing three Lidgerwoods, were so close to the metropolis of Seattle that the loggers could travel to and from work by streetcars. There the cutover lands were being cleared of stumps at an expense, we were told, of a hundred and fifty dollars an acre. I suppose that these lands by this time form a garden section of Seattle.

On our return trip in 1911 we had a chance to see the St. Joe National Forest in northern Idaho, or, rather, what was left of the forest after the terrible fire of 1910. Had it not been for the value of the fire-killed western white pine, which retains its good qualities for a few years, the Forest Service might as well have abandoned some four hundred thousand acres of the St. Joe Forest with its dead stumpage of two billion feet. As it was, the species other than white pine — Douglas fir, larch, white fir, and red cedar — were a total loss. The Milwaukee Lumber Company of St. Maries, Idaho, had the temerity to try the impossible and made of it a howling success. The white pine logs were brought over many miles to a standard-gauge railroad by horses, donkey engines, and Lidgerwood machines, and by chutes to a steel and concrete sawmill next to a lumberyard with a capacity of thirty million feet of white pine. The Biltmore foresters learned at the St. Joe National Forest that defeat could be turned into victory!

On another return trip from the West the Biltmore Forest School inspected the national forests near Salt Lake City, with their Engelmann's spruce, white fir, Douglas fir, and cottonwood, and with their recent afforestations — another new world for us.

Several Pacific Logging Congresses were taken in by the Biltmore Forest School, either on the way to the West or on the return to the East. These congresses, at which lumbermen, millmen, loggers, and foresters assembled annually, had been conceived in 1909 by my friend George M. Cornwall, editor of the *Timberman*. The high and the low, the competitors and the co-operators, the employers and the employed, the purveyors and the customers, and the rank and file of the woods came together for a few days of frolicking and exchanging of ideas. Whatever was new in machinery, in saws, and in other lumbering equipment was sure to be exhibited and discussed. Among the leaders present at such occasions in those years, let me mention a few: W. W. Peed of the Hammond Lumber Company of Eureka, California; J. Harold Bloedel and John J. Donovan of the Bloedel-Donovan Lumber Mills, Bellingham, Washington; H. C. Clair of the Twin Falls Logging Company, Portland, Oregon; T. P. Jones of the Potlatch Lumber Company, Boville, Idaho; R. W. Vinnedge of the North Bend Lumber Company, North Bend, Washington; J. D. Young of the Inman-Poulsen Logging Company, Kelso, Washington; B. R. Lewis of the Clear Lake Lumber Company, Clear Lake, Washington; and E. T. Allen, forester of the Western Forestry and Conservation Association. There were always instructive papers presented by these leaders and by leading foresters and logging engineers.

Every Pacific Logging Congress has set a milestone in the history of Western forests, Western industries, and Western manhood, ingenuity, and resourcefulness — which means to say a milestone in the progress of American forestry. For a teacher of forestry there never was a school better than a Western logging congress.

So far as I recall, the now historic Weeks Law, passed by the Congress in Washington in March, 1911, was not discussed at the Pacific Logging Congresses of those days. By this law the federal government authorized, for the first time in the history of American forestry, what it ought to have done a hundred years before, namely, the payment by the federal government of part of the cost of forest-fire protection incurred by the various states and owners of private forest lands. Perhaps the law had been passed too recently to awaken the interest of the lumbermen to its possibilities. Nevertheless, the Weeks Law was the first sign of good will shown by the United States Forest Service to the Western lumbermen, who had been paying taxes to their states and to the nation year after year without receiving that help, encouragement, and protection to which a taxpayer is entitled. Taxes are nothing but the price paid for governmental services needed by the taxpayer. No industry, no business, forestry included, is possible anywhere in this world that does not enjoy the good will and the protection of its government, and that fails to obtain from it in one way or another the equivalent of the taxes that it is compelled to pay.

The year 1911 and the Weeks Law marked the end of the infancy of American forestry. To recapitulate: The federal baby was born in 1881 when a Division of Forestry was created in the United States Department of Agriculture. Ten years later, in 1891, the first federal forest reserve was established. Another decade elapsed, and the year 1901 saw the youngster established in a national Bureau of Forestry. At the end of the fourth decade, under the auspices of the new Weeks Law in 1911, began the adolescence of forestry throughout the United States.

In this recapitulation nothing has been said about American forestry practiced on land privately owned. There was none, excepting that established at Biltmore in 1891. This first breaking of virgin soil, it is true, came to a close in 1909. But who will deny that, within the short span of its lifetime, the ground was effectively broken for private forestry to flourish in due course throughout America?

Naturally, for the students of the Biltmore Forest School, the visits and the excursions connected with its transcontinental tours were most instructive. While it was impossible for them to retain more than a fraction of the kaleidoscopic information registered on their brains in the course of their tours, they came in direct eye-to-eye contact with the possibilities and the impossibilities of American forestry, with the work of the United States and other countries, and thus were better fitted to meet the problems and hardships facing them in their coming careers. Important, too, it seems to me, is the fact that a teacher or a student confined merely to one site — for example, to Yale, to Duke, to Ann Arbor, to Berkeley, to Seattle — cannot help but consider the conditions existing at and near his alma mater as standard conditions. Is there any sense in forcing a forestry student to spend all his years of learning in one single narrowly-restricted set of laboratories, when there are hundreds of them available in the United States for the mere asking?

My views were not shared by the American fathers and mothers, nor by their sons whom I was eager to draw to the Biltmore Forest School. Its "odyssey," begun in 1909, came to an end with the close of 1913, when there were no new students in prospect. Retrospectively, let me assert that the Biltmore Forest School died at the right time. It died when it had reached the apex of its career. Be it man or tree or institution, it is better to die too early than too late.

CHAPTER 23

A Farewell Message to
My Biltmore Boys

DURING THE "ODYSSEY" of the Biltmore Forest School, a monthly newsletter, usually written by the class president, was prepared and mailed to all former students of the school. I knew that, in whatever fields of endeavor they were engaged, they would be interested in the happenings going on at their alma mater. The letter was called *Biltmore Doings*. The last issue, which bears the date of January 1, 1914, I wrote myself after the Biltmore Forest School had disbanded and I had returned to Germany. Here are some of the things I wrote:

This is the last issue of "Biltmore Doings." There will not be any more "Doings" to report on.

It was nineteen years ago that I left Germany for the United States of America.

And if I have now decided to leave in turn the United States of America for Germany, intending to spend the balance of my active life in the fatherland, my decision has not been prompted by any discontent over the events and happenings incident to those nineteen years of my American strife and life.

The best fortune I could have met with anywhere became mine in America: — fine fields to work in; good health to enjoy; enough to live on; and lots of friendship.

What more could my ambition have desired? And why,

then — you will ask — do I now abandon what I have nour-
ished and cherished for nineteen years? . . .

For the last four years, after parting with all active work
in the woods, I had centered all my nerve and sinew and
energy, and all my hopes, on the development of a really
American Forest School. That school was not meant to be an
institution of the usual kind, viz. a school attached to a college
away from the woods, a school preaching conservation and
second growth and theory. No! My "Biltmore Forest School"
was meant to be a practical and technical school, the teachings
of which, notably in lumbering and in financing, might be
capable of immediate application in the American woods; it
was to be a training school for the sons of every lumberman
and of every timber owner in the country.

It was to be; it has not been.

When in 1898 the Biltmore Forest School entered the arena
of American lumberdom, it was the only forest school in
America.

While its success did not break any records, it was successful enough in supporting itself.

Today, things have changed, or seem to have changed. Today, lumbering is supposed to be taught, in one way or
another, at no less than eighty-three American schools.[1] There
seems to be no more need, therefore, of a unique school like
the Biltmore Forest School. As a matter of fact, the enroll-
ments at the Biltmore Forest School have been so small, recently, that its continuance is not worth while — is not in
keeping with the dignity of the interests which I have advocated incessantly.

The Biltmore Forest School has had nothing to support it
except the goodwill, free from any financial obligations, first
of George W. Vanderbilt's Biltmore Estate, and thereafter of
such firms as Cummer-Diggins at Cadillac, Mich.; of C. A.
Smith, at Marshfield, Ore.; of J. L. Roper, at Norfolk, Va.
To these firms we are indebted deeply for their copious hospitality and for their encouraging friendship. Had we had
fifty such friends . . . this Biltmore Forest School might
have been "conserved."

In the fall of the past year, by addressing all leading lum-

---

[1] At that time there were in the United States twenty-two forestry schools,
including Biltmore.

bermen known to me in the country, I have attempted to enlist more of an active interest in my Biltmore Forest School, in its ideas, in its purpose, in its work. Mind you: I did not seek to get their money; not at all. I asked for a word of encouragement, and for students to be taught, hoping that by the answers I might prove to myself that my work was wanted.

The proof was not forthcoming; and less than 20 students are now enrolled at this school. That is a small class of students; and I am not willing to teach a small class for two reasons: first, because I cannot teach without the inspiration of a larger audience; second, because I cannot foot the bill of a staff of teachers on fewer than some forty students.

The conclusion is evident that such a school as I had been planning or as I had been developing, is not so badly needed by the lumber interests of the United States, as I had been supposing to be the case.

. . . .

Nevertheless, I trust and I pray, that, though the school be dead, my graduates and also my friends among the American lumbermen will no more forget the Biltmore Forest School and its teachings than I can forget them. When they, — when you come to visit the Old Country you will find me welcoming you at Darmstadt . . . where I shall hold some "job," trying to graft some American ideas on German forest work.

And when I come to America to visit, as I shall do certainly from time to time, I hope to have you extend to me the same glad hand that I have shaken in the past.

. . . .

Here, perhaps, I ought to stop. But a natural solicitude for that visionary truly American Forestry to which the best years of my life have been devoted, and a keen desire to epitomize that devotion, cause me to add a word of loving advice to a word of loving farewell.

We should not expect that the forests will be conserved, in the long run, by the Nation or the States or the owners, unless it pays to conserve them. It cannot pay to conserve them, unless the price of stumpage soars high, and rises to soar still higher.

You, the consumer, must pay the bill of higher prices; and

I beg of you to pay it willingly for the benefit of the country which we all love together.

We should not attack and vilify those who have made a success from timely investments in timberlands. They are — in the same manner as successful farmers or miners or brokers — reaping the reward of farsightedness. The very rarity of the successful timber owner in a field where the opportunities were endless, is proof of his peculiar talents. And because a genius deserves a reward, I beg you to bear no grudge against those who had the talent or the genius which you or I either did not possess, or did not apply.

We should not blame the man who transforms the primeval forests into barren wastes; you and I, in his place, would certainly act as he does. Virgin woods are utterly unproductive, just as much so as the barren wastes that take their places. The conversion of the primeval forest directly into the cultured forest is too difficult a task, and too risky a venture for anyone to attempt: — But, if we do attempt it, let us entrust the task to the logger who knows some "forestry" rather than to the forester who knows some "logging"! — Forestry resulting in a second growth must come by slow evolution and from the willing efforts of those within the logging camp; it cannot come by a quick revolution and by pressure of public opinion from without. The same principles of evolution apply to lumbering, in other words, as they do to farming and mining and business in general. The possibilities of a second growth depend on its profitableness and on the safety of the investment which it represents. These very factors have been influenced negatively by our thoughtlessness in regard to forest taxation, and by our laxness in regard to forest fires.

. . . .

I grant that even to-day, here and there, under extraordinary conditions, a profitable conversion is possible of the old forest directly into the new; and I grant gladly that this conversion will become possible more and more, as time goes on, and as those extraordinary conditions — conditions which put the financial burden of the transaction on the consumer of stumpage — arise in more and more localities. Still the natural permanency of our forests is and will be based on the permanent financial success of their owners, and that in turn depends on wise legislation born from the well wishing good-will of an enlightened public opinion. To the end that such kindly inclination towards the owners might be secured, we require an organization all over the United States representing every owner of woodlands. The organization should be formed for the distinct purpose of acquainting the American public with forestry as an American business possibility. The perpetuation . . . of our forests when they stock on a soil unfit in the near future for successful farming . . . is a question of financing; some body must be willing to finance that famous "forest conservation." Who is it to be?

If we in America fail to render fairly remunerative on the one hand, and on the other fairly safe, such investments as might be made in conservative forestry, the permanent maintenance of private forests in America on a scale larger than woodlots is made impossible.

The case of the forests is no other than that of the railroads.

Our railroad system must cease to be maintained properly, just as soon as private railroad investments cease . . . to be remunerative and safe. The same logic applies to our forests privately owned.

And this similarity, or this parallelism of railroads and forests — so it seems to me — goes still further:

If our railroads — and our private forests, too — owing to financial debacles due to one cause or another, fail to be maintained properly, then either the Government or else the States are forced in the end to step in, as owners, and to run the railroad and the forest business on their account, — that is, the public's; and that means or implies a stride not towards but into a socialism absolutely opposed to our traditions.

The railroads or the forests may yield profits under public ownership provided that the consumers will pay the bill, in one way or another.

If the railroads or the forests should fail to yield profits, if they are run and maintained at a loss by the Government or by the States, then the whole nation must be taxed to shoulder the deficit.

There is no escape: — Pay we must, either with the right hand or with the left, for every benefit that we obtain from the railroads or from the forests, either as consumers or else as a nation.

It is well known, I suppose, that 200 million acres of our American forests (out of a total of 500 million acres) are already owned or held, socialistically, by the United States of America and by a few of the States, while another 300 million acres of forests are owned privately.

The private forests contain the most and the best timber; but they are not maintained because they cannot be maintained at a profit. The State and the National Forests are maintained; but they are maintained at a loss; and the people pay the tax to meet the annual deficit.

Nobody claims, as far as I know, that there is no need to maintain the forests: But the question has arisen frequently, and will arise again: — Is there any need to maintain this or that distinct forest, or to maintain that 500 million acres of forests which we now possess in America? Where is the limit to our deforestation?

My answer to the question would be this: —

Where the forest occupies, in large tracts and in a suitable climate, a soil actually fit for remunerative and continuous farming, there the forest *must* disappear as soon as, — but no sooner than —, the expanding interests of American farming require the additional soil.

The limit of coming deforestation is the fence of the coming farm.

If that is so, then it is of the utmost importance that the terms "farm-soil" and "forest-soil" should be given a legal interpretation; that the realm of the farm and the realm of the forest should be delineated everywhere, definitely and topographically; that our forest laws should differentiate strictly between the forests stocking on actual farm-soil, and all other forests.

I have said that our National Forests are being maintained at a loss, and that the people are being taxed to meet the annual deficit. The uninitiated is apt to suspect that the blame belongs with the National Forest Service whereas the fault lies with the conditions. These conditions fail to force the consumer to pay for the commodities produced by the National Forests that price at which these commodities can be reproduced: The cost of reproduction is greater by far than the present price of the products.

Thus it is the conditions of price that are at fault, not the

Forest Service when the National Forests are maintained at a loss.

. . . .

I have known the staff of the National Forest Service for many years; many of its members have had their training at the Biltmore Forest School; public servants more faithful than those of the National Forest Service do not exist anywhere, in any country. It is true, of course, that mistakes are being made by some of them, here and there, now and then. But the good will and the good intentions and the high ideals of all National Forest Officers can not be questioned for a moment. They have been subjected, nevertheless, to the most severe criticism. We are bound to know, from our own business organizations, the disastrous effect of harsh and unjust and sweeping vituperation: The critics, if they actually care to improve the National Forests and the National Forest Service, should remember that ancient truth which applies with great force to our work in American forestry: — *in magnis et voluisse sat est.*

Good bye! God bless you, and the United States of America, and all the workers in her forests! . . .

Fondly in Biltmore yours
C. A. SCHENCK

So ran my message and my thoughts about American forestry in January, 1917. Later in that same year occurred two events of fateful consequence to Pisgah Forest. George W. Vanderbilt died in March from an appendicitis operation in Washington, D.C., and towards the end of the year Mrs. Vanderbilt negotiated a sale of Pisgah Forest to the United States government. It was characteristic of her to want the forestry dreams of her husband and the beautiful mountain country in which they had taken root saved, not merely as an object lesson, but also for the use and enjoyment of the American public. She might have sold Pisgah Forest in parcel lots to lumbermen and speculators at prices that in the end would have aggregated a large sum. Instead, she turned over upwards of a hundred thousand acres to the National Forest

Reservation Commission at a price far below what she might have obtained, specifying merely that the area should be embraced in a national forest to be known as the Pisgah National Forest.

Thus was my beloved Pisgah Forest saved from dissection and destruction. During the almost forty years that have elapsed since 1914, it has been under the protection and management of the United States Forest Service. Of course there is sorrow for me in the thought that it has failed to mature as a private enterprise, as George Vanderbilt and I had so bravely planned. Nevertheless, I am thankful, and I pay everlasting tribute to Mrs. Vanderbilt for having preserved those forested scenes of our struggles, so that the millions of Americans, be they foresters or not, who through the years frequent their hills and vales may draw inspiration from their beauty and some lessons from our pioneer forestry work, now mellowed by half a century of time.

And now let me conclude these recollections of early forestry in America with a creed which I found, not on a Sunday in church, but on a week day in the office of a lumberman, eighty-nine years old and incidentally the oldest graduate — an honorary graduate — of the Biltmore Forest School. He is J. H. Bloedel of Seattle, Washington, and his creed may be summarized in these words: To be active, to be proud, to be tolerant, to have faith, to have integrity, and to be grateful.

Grateful! Gratitude is the best of all qualities; and God knows, I myself have so much cause to be grateful that I feel I should be walking on my knees rather than on my feet.

I am grateful that I have had the very best of all mothers and a beloved mate; both of them have been an inspiration to me while they lived — and ever since!

I am grateful for the vision of high purpose of the founders of this republic in establishing a system of competitive and free enterprise. Under this system and during my lifetime, I have seen the United States grow from a country of sixty-five million to one hundred and fifty million people, enjoying freedom, comforts, and opportunities utterly unknown to less fortunate peoples. I am grateful that I have had a small part in developing and preserving that system.

I am grateful in particular for the continued love and lasting attachment of my boys of the Biltmore Forest School, which, though it came to an end forty years ago, continues to live in the hearts of a large number of sterling Americans — yea, even in those of their progeny.

Some Publications
by the Author

THE CAPITALIST and Economic Forestry: An Address Read
at the Special Meeting Held by the American Forestry
Association on August 22 and 23, 1899, at Columbus,
Ohio. Asheville, North Carolina, n.d. [6 p.]

The Problem of Forestry in Minnesota: Report to the State
Forestry Board of Minnesota. St. Paul, 1900. 11 p. Also
printed in the Fifth Annual Report of the Chief Fire
Warden of Minnesota, 1899, p. 125–136.

Forestry as Applied to Reservations Used as Parks. n.p., n.d.
10 p.

Forestry for Kentucky: A Stereopticon-Lecture Delivered at
the Invitation of the Louisville Board of Trade by C. A.
Schenck, Ph.D., Forester to the Biltmore Estate. n.p., n.d.
[8 p.]

Financial Results of Forestry at Biltmore: Address Delivered
by Dr. C. A. Schenck at the Summer Meeting of the American Forestry Association at Minneapolis, August 26, 1903.
n.p., n.d. 6 p.

"The Forestry Interests of the South." In the Tradesman,
20th annual, p. 95–99, January 1, 1899.

Forestry and Forest Reserves: An Address Delivered at the
Second Annual Meeting of the Hardwood Manufacturers'

Association of the United States on January 26, 1904. n.p., n.d. 6 p.

Compound Interest Tables Prepared for the Use of the Students of Biltmore Forest School. n.p., n.d. [12 p.]

Biltmore Forest: A Forest Fair in the Biltmore Forest. Asheville, North Carolina, November 26, 1908. 56 p.

Cruisers' Tables Giving the Contents of Sound Trees, and Their Dependence on Diameter, Number of Logs in the Tree, Taper of Tree and Efficiency of Mill. n.p., [1909]. 62 p.

Lectures on Forest Policy: Forest Policy Preamble. Asheville, North Carolina, 1903. 48 p. Second Part: Forestry Conditions in the United States. [Asheville, North Carolina, 1904]. 108 p.

Forest Management. [Asheville, North Carolina, 1905]. 16 p.

Forest Management (Forest Working Plans): Guide to Lectures Delivered at the Biltmore Forest School. Asheville, North Carolina, 1907. 34 p.

Forest Mensuration. Sewanee, Tennessee, 1905. 72 p.

Biltmore Lectures on Silviculture. Albany, New York, 1905. 182 p.

Forest Finance. Asheville, North Carolina, 1909. 44 p.

Forest Protection: Guide to Lectures Delivered at the Biltmore Forest School. Asheville, North Carolina, 1909. 160 p.

Forest Policy. Darmstadt, Germany, 1911. 168 p.

The Art of the Second Growth, or American Silviculture. Albany, New York, 1912. 206 p.

Forest Utilization. n.p., [1904]. 118 p.

Logging and Lumbering, or Forest Utilization: A Textbook for Forest Schools. Darmstadt, Germany, [1912]. 190 p.

National Rules of Hardwood Inspection Effective December 1, 1907: Compiled for the Biltmore Forest School. Asheville, North Carolina, [1907]. [6 p.]

In the Woods of Minnesota. Chicago, [1896]. [8 p.].

Index

PLACES NOT LOCATED BY STATE OR COUNTY ARE IN NORTH CAROLINA.

ADAMS, GEORGE L., tannic-acid plant, 103, 149
Adirondack Mountains, forests and forestry, 43, 87, 88, 93, 106–110, 185, 186
Afforestation, Biltmore: 56, 57, 162, 163, methods, 45, 46, results, 55; Pisgah Forest, 49, 50, 97, 163; Coxehill, 163. *See also* Forestry
Ahern, George P., visits Biltmore, 125
Alabama, longleaf pine study, 86
Allen, E. T., at logging congress, 198
Allison, H. O., Biltmore lectures, 152, 190
American Can Co., box factories, 187
American Forest History Foundation, 176
American Forestry Assn., 18, 142
American Lumberman, editor, 164
American Museum of Natural History, Jesup collection, 19
Andrews, Christopher C., sec. Minnesota forestry board, 72; state fire warden, 72–74
Appalachian parks, 122; movement for establishment, 118; Schenck's proposal, 119–122; Scott's views on, 157
Archbold, John D., visits Biltmore, 157
Arnhem (Holland), Biltmore school visits, 179
Arnold Arboretum (Jamaica Plain, Mass.), director, 22

Asheville, 1, 157; hotel, 23, 27, 146; furniture factory, 24; sawmill, 24, 41, 44, 49; market for Biltmore wood, 26, 47, 60, 61, 126; Roosevelt's speech, 112; number of visitors, 133; saloons, 152; social life, 167; Eliot's lecture, 169; Biltmore school headquarters, 178
Asheville and Chicago Hunting Club, leases Pisgah rights, 172
Austria, forests, 154
Avery's Creek, 51, 64; store, 134
Axton (N.Y.), 109; nurseries, 107, 186

BACKUS, GORDON T., forestry student, 80; work in Highland Forest, 114
Baden (Germany), forests and forestry, 13, 182; Brandis' notes on, 14
Baird, J. H., editor, 124
Bavaria (Germany), Brandis' notes on forests, 14
Beadle, Chauncey D., heads landscape dept., 24, 170; encounter with Schenck, 171, 172
Bell, Alexander Graham, invention of telephone, 138
Bellingham (Wash.), lumber mills, 198
Berckheim, Count, arboretum, 181
Big Creek, logging, 28, 40, 49, 50, 53; valley described, 29; splash dams, 29, 35, 37, 40–42, 62; results of splashings, 42, 53. *See also* Pisgah Forest

213

Biltmore (village), 162; described, 23, 126, 127; church, 31, 127; woodcutting plant, 60

Biltmore Arboretum, 25, 36, 170

Biltmore Estate, described, 1, 23–26; Pinchot's forestry plan, 18; map, 23; brick factory, 24; farms, 24, 25, 27, 163; summer homes, 25; logging, 25, 46; acreage, 25, 51, 51n.; roads, 46, 75, 90; fish hatcheries, 51, 60; afforestation, 55–57, 162, 163; forest nurseries, 56, 57, *see also* Biltmore Nurseries; pheasant hatchery, 60; interior holdings bought, 91; shops, 100; hospital, 137; forest fires, 143, 148; tree species, 162, 163; herbarium, 163. *See also* Biltmore Forest; Farm department; Forest department; Landscape department; Pisgah Forest

Biltmore Forest, 13, 14; fires, 26; tree species, 26, 45, 101; described, 28; map, 33; roads, 43; logging, 44, 167; afforestation methods, 45, 46; rangers, 51; Schenck's accomplishments, 175. *See also* Biltmore Estate; Forest department

Biltmore Forest Festival, 161–165

Biltmore Forest School, 58, 61, 80, 151; U.S. tours, 38, 185–200; established, 77, 79, 144; studies and lectures, 78, 90–92, 118, 126, 137, 190; Alabama pine study, 86; guest lecturers, 91, 92, 138, 152, 165; European tours, 94, 95, 126, 130, 132, 154, 175, 178–185; enrollment, 97, 126, 202, 203; graduates assist Schenck, 114–116; building, 126; expenses, 126; closing asked, 132; clubhouse, 152; sängerfests, 152; 168; working fields, 156, 171, 172, 174; 10th anniversary celebration, 161, 162; self-supporting, 173; effects of Schenck's dismissal, 173, 174; textbooks, 176; advantages to hosts, 190; cruising methods, 190; closes, 200; Schenck's farewell message, 201–207

Biltmore House, 24, 36, 79, 131, 168; architect, 23, 98; described, 32; guests, 98–101, 112, 156; closed, 133; concert, 137

Biltmore Lumber Co., losses, 68

Biltmore Nurseries, 24, 36, 56, 57, 118, 163; tree species, 25, 36

"Biltmore stick," invented, 81

Black, Frank S., gov. New York, 93

Black Forest (Germany), 3, 6, 179; forestry, 53; described, 182

Bloedel, J. Harold, 198, 208

Bloedel-Donovan Lumber Mills, 198

Bonn (Germany), 5, 8

Booth, John R., white pine holdings, 145

Boville (Idaho), lumber co., 198

Brandis, Sir Dietrich, 22, 40, 55, 80, 92; heads European forest tours, 4–7, 10, 11, 183, retires, 16; inspector general of forestry in India, 4, 10; influence on American forestry, 8; biographical sketch, 8, 9; Pegu forest supt., 9, 10; advises Schenck, 11–14, 41; author, 11, 14, 15; recommends Schenck, 17, 18; influence on Schenck, 78; death, 154

Brandley, Frederick, landscape foreman, 62

Brenner, Fred, lumber agent, 153

Brevard, 65, 66; tannic-acid plant, 103, 104

Brick Farm House, 31; proposed clubhouse, 146

Brimley, C. S., Biltmore lectures, 152

Broadhurst, Edgar D., Biltmore lectures, 152

Bronson, Daniel D., Biltmore graduate, 133

Brooklyn Cooperage Co., contract with Fernow, 106–109; logging, 186

Bryan, William Jennings, presidential candidate, 33, 161

Bryce, Viscount James, visits Biltmore, 100, 101

Buckspring, hunting lodge, 58, 59, 99, 100, 127

Burton, Ralph G., assists Schenck, 116

CADILLAC (Mich.), Biltmore school headquarters, 178, 189–192

Cambridge University, forestry instruction, 5

Camp Perry, lumbering, 186

Canada, forestry, 139, 140, 145; gov. general, 144, 145, 156, 157; dominion nurseries, 146

Canadian Forestry Congress, *1906*, 144–146

Cannon, Joseph G., on Minnesota excursion, 89

Canton, pulp wood factory, 148; fibre co., 187

Carr, Louis, lumberman, 68n.
Carr Lumber Co., 68n.
Cary, Austin, influenced by Brandis, 8; visits Biltmore, 125
Case, James, ranger, 134, 135
Cass Lake (Minn.), island named, 89
Catawba River, log boom, 37
Cecil, George H., forester, 193
Champion Coated Paper Co., 148, 189
Champion Fibre Co., 189; buys Pigeon River tract, 52; logging and lumbering, 187, 188
Chantilly (France), forests, 184
Chapman Timber Co., 193, 194
Cherokee Indian reservation, logging, 39
Chippewa Indian reservations, efforts to convert to forest preserves, 88–90
Chippewa National Forest, Biltmore school visits, 193
Cincinnati (O.), veneer mills, 189
Clackamas (Ore.), nurseries, 194
Clair, H. C., at logging congresses, 198
Clark, Greenleaf, vice-pres. forestry board, 75
Clear, Lake (N.Y.), nursery, 186
Clear Lake Lumber Co., 198
Cleveland, Grover, U.S. pres., 43; asks appointment of forest commission, 47; proclaims forest reserves, 48
Cleveland Cliffs Iron Co., properties, 158, 159
Clyde Iron Works, 38, 193
Cobb, Collier, Biltmore lectures, 92
Coeur d'Alene National Forest, Biltmore school visits, 194
Cogburn, John, cabin builder, 37
Columbia River, forests, 193
Condon, C. D., Biltmore lectures, 152
Cooper's Hill (England), engineering college, 4, 5, 7, 16
Coos Bay (Ore.), Biltmore school visits, 178; sawmill, 194
Coos River (Ore.), lumbering, 195, 196
Cornell University, 108; plans forestry course, 81; forestry school, see New York State College of Forestry
Cornwall, George M., editor, 193, 198
Corundum mine, Highland Forest, 113
Couvet (Switzerland), forests, 184
Coxehill, afforestation, 163

Cross, Judson C., pres. forestry board, 75
Crown Columbia Paper Co., properties, 194
Cummer-Diggins Lumber Co., Biltmore school host, 189–192, 202

DALHOUSIE, JAMES RAMSEY, EARL OF gov. general of India, 9
D'Allinge, Baron Eugene, heads farm dept., 24, 33
Darmstadt (Germany), 93, 132; chemical firm, 19; Schenck's home, 153, 203; Biltmore school headquarters, 178, 179; forests near, 180–182
Darmstadt Institute of Technology, Schenck at, 2, 4; Biltmore school affiliation, 174, 177, 180
Davidson River, 64, 66; forest conditions, 30; tanbark and tannic wood, 104, 105; camp, 127, 128; road, 129; pheasant hatchery, 137
Deer park, Biltmore, 59
Defebaugh, James E., editor, 164
Diggins, Fred, lumberman, 191
Donovan, John J., at logging congresses, 198
D'Osmoy, Viscount and Lady Romain, 100
Dresser, D. Le Roy (Roy), pres. shipbuilding co., 128
Dryden, John F., U.S. senator, 146
Ducktown (Tenn.), copper mines and smelters, 114
Duluth (Minn.), Biltmore school visits, 38, 193
Dürer, Albrecht, prints collected by Vanderbilt, 33

EASTERN AND WESTERN LUMBER CO., mills, 194
Edison, Thomas A., interest in radium, 71
Eliot, Charles W., visits Biltmore, 168, 169
Elliott, F. A., Oregon state forester, 193
Erzgebirge Forest (Germany), 14
Eureka (Cal.), lumber co., 198

FAN, WU TING, ambassador to U.S., 99
Fankhauser, Franz, forester, 11
Farm department, Biltmore, 57, 92
Fernow, Bernhard, 23, 47, 72, 87, 92; sec. forestry congress, 21; biograph-

ical sketch, 21; relations with Schenck, 21, 22, 109, 110, with Pinchot and Sargent, 109; heads U.S. Div. of Forestry, 21, 78; visits Biltmore, 39; efforts to consolidate U.S. forestry movement, 61; speaker, 62; heads Cornell school, 91, 93, 106; Adirondacks forestry, 93, 106–110; contract with cooperage co., 106–109; forestry dean at Toronto, 146, 151, 160. See also New York State College of Forestry

FERNOW, BERNHARD, Economics of Forestry, 110

Fernow, Mrs. Bernhard, 22, 61

Financial depression, 1907, effects on forest dept., 153–155, 159

Fish hatcheries, Biltmore, 51, 60

Fontainebleau (France), forests, 184

Ford, Paul Leicester, visits Biltmore, 99

Forest department, Biltmore, 160; lumber and fuel wood business, 25, 44, 47, 60, 61, 90; financial control, 35; lawsuits, 90, 91; tannic-acid plant contract, 103–106, 129; real estate sales and hunting leases, 127, 172; financial difficulties, 133, 147, 153–155, 159, 167, 171, 172; stores owned, 134; pine seedling sales, 170. See also Biltmore Estate; Biltmore Forest; Pisgah Forest

Forest fires, Biltmore, 26, 97, 143, 148, 171; Nehasane Park, 88; St. Joe National Forest, 197; protection against 199

Forest Leaves, editor, 22

Forest owners, number in U.S., 166; need for organization, 205. See also Forestry, private

Forest rangers, Biltmore: 25, 29, 30, 51, 59, 60, 63, 64, 134, 135, salaries, 121, uniforms, 128, 129; Germany and France, 119

Foresters, lack of trained, 16; attitude toward lumbering, 136; English-Indian, visit Biltmore, 138

Forestry, Brandis' influence on American, 8; India, 9–12; Indonesia, 10; Pennsylvania, 13, 22; Germany, 13, 53, 179–183; relation to transportation, 34, to logging and lumbering, 53, 54, 70, 76, 93, 110, 118, 139, 143, 144, 175, 202, to silviculture, 133; as an investment, 43, 44, 67, 68; Adirondacks, 43, 87, 88, 93, 106–110, 185, 186; importance of roads,

47, 67; Sweden, 72; legal principles, 91; at Biltmore, 97, 98, 133, 175, 176, 186–189, see also Biltmore Estate, Biltmore Forest, Pisgah Forest; importance of Roosevelt-Pinchot relationship, 111; conservative, in U.S., 121–123; private: 121, 125, Schlich's views, 11, Brandis' views, 11–14, working plans, 12–14, impediments, 142, subdivisions, 176, in U.S., 200; public: 121, propaganda for, 140–142, growth in U.S., 141, 199, subdivisions, 176; as a science, 139; Canada, 139, 140; growth in U.S., 141, 199; influence on runoff, 158; importance of lumber tariff, 166; Holland, 179; Switzerland, 183, 184; France, 184, 185; Michigan, 189–192; U.S. West, 192–197. See also Afforestation; Silviculture

Forestry apprentices, Biltmore, 43, 51, 61, 77

Forests, national: establishment recommended, 47, controversy, 48, Hawaii, 141, establishment of first, 199, maintained at a loss, 206, 207; acreage of public, in U.S., 206. See also Forestry; individual forests and localities

Fort Wayne (Ind.), catalpa plantations, 189

Fosdick, Dr. Harry E., quoted, 124

France, Brandis' tours, 5; forest rangers, 119; Biltmore tours, 184, 185

Frankfurt (Germany), forests, 181

French Broad River, 28, 29, 42, 104, 127; plowlands, 26; sawmill, 34; log boom, 38, 39, 41; logging, 44, 49; proposed golf course, 146; floods, 158

GADSKY, JOHANNA, Biltmore concert, 137

Gaildorf (Germany), forests, 183

Gannett, Henry, opinion on timber canvass, 117

Garden and Forest, 43; editor, 14, 22, 48

Gaskill, Alfred, forestry student, 80

Georgia, state sues Tennessee, 114; forest fire laws, 115

German Forest Service, 77, 93

Germany, Brandis' tours, 5, 10; ambassadors to U.S., 101, 102; Biltmore tours, 179–183

Giessen (Germany), forests, 5

Gillespie, George, ranger, 60

Glacier National Park, Biltmore school visits, 193
Graham, F. W. W., 127
Grand Rapids (Mich.), Biltmore school visits, 189
Graves, Henry S., influenced by Brandis, 8; author, 31n.; in Rockies, 47; Nehasane forestry, 87, 88, 92; Biltmore experiences, 92; teaches at Yale, 92; U.S. chief forester, 186
Grays Harbor (Wash.), logging, 196
Great Northern R.R. Co., sponsors forestry excursion, 88–90
Greeley, William B., asst. U.S. forester, 186
Green, Samuel B., forestry professor, 72, 73
Grey, Albert Henry, Earl, gov. general of Canada, 144, 145; visits Biltmore, 156, 157
Grey, Lady Alice, 156
Griffith, Edwin M., influenced by Brandis, 8; forestry apprentice, 43; member foresters' society, 139
Griggs, Everett G., pres. lumber co., 196

Hall, William L., 186
Halstenbek (Germany), nurseries, 55, 56
Hamilton (O.), paper co., 148
Hammond Lumber Co., 198
Harding, Edward J., Biltmore controller, 137
Hardwood Dealers' Assn., 135
Hardwood Manufacturers' Assn., 135
Harvard University, forestry school, 168; pres., 168, 169
Havemeyer Sugar Co., 106
Hawaiian Islands, Div. of Forestry, 141
Hays, J. F., railroad pres., 103
Haywood County, road, 59
Heidelberg (Germany), forests, 181
Hendersonville and Brevard R.R., pres., 103; extended, 134
Herty, Charles H., visits Biltmore, 138
Hess, Richard, forestry dean, 3
Hickory, splash dam and sawmill, 37
Highland Forest, Schenck's work as forester, 113, 167, 189, 194
Holleben, Theodore von, ambassador to U.S., 21, 101
Hopkins, A. D., Biltmore lectures, 152
Hosmer, Ralph S., heads Hawaiian forestry div., 141

House, Homer D., Biltmore asst. forester, 160; Biltmore lectures, 190
Howe, Clifton D., Biltmore asst. forester, 151; forestry dean at Toronto, 151, 152, 160
Hubbard, Elbert, on forestry excursion, 89
Huchenfeld (Germany), forests, 183
Hunt, Richard M., architect, 23, 98, 99
Hurricanes, effects on forests, 192

Imperial Indian Forest Service, candidates, 5; established, 10
India, forestry, 9–12
Indian Head (Sask., Can.), nurseries, 146
Indonesia, forestry, 10
Inman-Poulsen Logging Co., 198

Jackson County, logging, 38
Jesup, Morris K., American woods collection, 14
Jones, T. P., at logging congresses, 198
Judeich, Johann F., forester, 11
Judson, William B., editor, 75

Kapowsin (Wash.), logging, 196
Kellogg, Royal S., Biltmore lectures, 190
Kelso (Wash.), logging co., 198
Kenilworth Inn (Asheville), 23, 27, 146
Kentucky, efforts for forestry assn., 123
Kern, Richard, Biltmore asst. forester, 179
Kern, Mrs. Richard, hurricane experience, 191
Ketchum, Mrs. Eleanor G., sec. to Schenck, 43
King, Hiram, carpenter, 63
Kitchin, William, gov. North Carolina, 176
Kuser, John L., 127

Lacey, James D., cruising firm, 194
Lafon, John, Highland forester, 114, 194
Lagerstroem, Cornell, 194
Lamb, Frank H., logging operations, 196
Land laws, federal, abuses, 140
Landscape department, Biltmore, 24, 61, 170
Langenbrand (Germany), forests, 183

Laurier, Sir Wilfred, Canadian prime minister, 145
Leupp, Francis E., editor, 40
Leupp, Graham, Biltmore student, 40
Lewis, B. R., at logging congresses, 198
Lewisohn (Adolph), and Sons, 114
Lidgerwood aerial skidder, 196, 197
Lindenfels (Germany), 2, 177, 179
Logging and lumbering, Brandis' views on, 12; at Biltmore: 25, 28, 29, 37, 40–42, 44, 46, 47, 49, 50, 53, 60, 90, 129, 167, taught, 137; relation to forestry, 53, 54, 70, 76, 93, 110, 118, 139, 143, 144, 175, 204; foresters' attitude toward, 136; Adirondacks, 186; Pigeon River, 187; Michigan, 190; U.S. West, 193, 195–198
Longworth, Mrs. Nicholas (Alice Roosevelt), visits Biltmore, 100
Lookingglass Creek, sawmill, 63
Lookingglass Rock, 68, 128
Lorey, Tuisko, forestry dean, Tübingen, 2
Lotbinière, Sir E. G. Joly de, 139
Lotbinière, Sir Henri Joly de, 139
Louisville (Ky.), Schenck's lecture, 123
Lumber, prices, 26; inspection rules, 135, 136
Lumbering. See Logging and
Lumbermen, attitude toward forestry, 13, 75, 97, 132; purchases in South, 53; attacked by Pinchot and Roosevelt, 132
Luther, Hans, ambassador to U.S., 102

McCall, Ananias, farmer, 63
McDonald, Thomas J., aids Schenck, 116
McDonald and Vaughan, lumber co., 196
McGiffert, John R., log loader, 38; skidder, 193
McKinley, William, assassination, 111
McKinley tariff, effects, 121
McNamee, Charles, Biltmore manager, 20, 23, 34, 35, 41; appoints timber sales agent, 49; relations with Schenck, 68; leaves Biltmore, 131
Malley, Jack, builds dam, 41, 42
Mannheim (Germany), chemical works, 115

Marshfield (Ore.), Biltmore school headquarters, 178, 194–196
Mather, William G., pres. iron co., 158, 159
Mayr, Heinrich, forestry professor, 92, 96, 158
Medora (N.D.), Roosevelt at, 111
Meister, Ulrich, forester, 11, 158
Merck, E., chemical firm, 19
Merck, George, chemical co. agent, 18, 19, 71
Merck, George W., Jr., director forestry assn., 18; pres. chemical co., 19
Mereen, Arno, vice-pres. lumber co., 194
Mereen, John, inventions, 195
Merrill and Ring Logging Co., 197
Michigan, lumbering, 190
Michigan Hardwood Manufacturers' Assn., pres., 189
Milford (Pa.), Pinchot's home, 93
Millard, C. I., lumberman, 186
Mills River, logging, 49
Milwaukee Lumber Co., 198
Mimizan-les-Bains (France), Biltmore school headquarters, 178; forests, 184, 185; naval stores industry, 185
Minneapolis (Minn.), lumbermen's meeting, 132, 135
Minnesota, forestry on school lands advocated, 72, 73; cutover lands, 72, 74; forest fire law, 74; forestry excursion, 88–90
Minnesota Historical Society, headquarters forest history foundation, 177
Minnesota Horticultural Society, 72
Minnesota State Agricultural Society, 72
Minnesota State Fish and Game Commission, 72
Minnesota State Forestry Assn., 72
Minnesota State Forestry Board, meeting, 72; Schenck's report, 75
Mohr, Charles, report on Southern pines, 87
Montreal (Can.), forestry meeting, 139
Moonshiners, Pisgah Forest, 64–66
Moore, Edgar B., heads sports club committee, 146
Mormons, aid mountaineers, 64
Morton, J. Sterling, U.S. sec. agriculture, 22; visits Biltmore, 27

Mount Rainier National Park, Biltmore school visits, 194
Mountaineers, Pisgah Forest, 30, 58, 63–67, 172
Muir, John, visits Biltmore, 37
MUIR, JOHN, *The Mountains of California*, 37
Müller-Alewyn, Max, Schenck's uncle, 16, 17
Murphy, Franklin, gov. New Jersey, 147
Murrill, William A., visits Biltmore, 138; consulted on chestnut disease, 160

NASHVILLE (Tenn.), Schenck's lecture, 123
National Academy of Sciences, names forest commission, 47
National Conservation Commission, forest committee, visits Biltmore, 157, 157n., 158
National Forest Commission, sec., 31n.; recommendations, 47
National Forest Reservation Commission, 207
National forest reserves. *See* Forests, national
National Lumber Manufacturers' Assn., meetings, 132, 165; pres., 196
National parks, movement to establish in Appalachians, 118
Naval stores industry, France, 185; North Carolina, 185, 186
Nebraska, Arbor Day movement, 22
Nederlandische Heide Maatchappij, (Holland), forestry, 175, 179
Nehasane Park (Adirondacks), forestry, 43, 87, 88, 93; fire, 88
New Bern, forestry meeting, 62; Biltmore school headquarters, 178, 186; mills and factories, 187
New York (N.Y.), Schenck's impressions, 19, 20
New York Central R.R., decline of stock, 133
New York State College of Forestry (Cornell University), 97, 109; headed by Fernow, 90, 91; Adirondacks working field: 93, 106–110, court forbids lumbering, 108; state appropriations withdrawn, 108; library, 176. *See also* Cornell University
New York State Commission of Forests, Fish, and Game, 106

Newell, Frederick H., Biltmore lectures, 92
Norfolk, shipping, 187
North Bend (Ore.), mills and shipyards, 196
North Bend Lumber Co., 198
North Carolina, acquires new factories, 103; forestry conditions, 142; pine canvass, 150; state entomologist, geologist, 152; prohibition, 152, 153; forestry laws, 176; state forests established, 176; naval stores industry, 176, 185, 186
North Carolina Geological Survey, discourages gold search, 70
Northwestern Lumberman, editor, 75; articles by Gannett and Schenck, 117
Norwood Manufacturing Co., 186

OBERHOLSER, HARRY D., Biltmore lectures, 152
Ohio State University, Schenck's lecture, 87
Olmsted, Frederick E., influenced by Brandis, 8; member foresters' society, 139
Olmsted, Frederick L., landscape architect, 23, 25, 33; Schenck's host, 24, 31, 36; portrait, 98; recommends Beadle, 170
Olmsted, Mrs. Frederick L., 24, 25, 27
Olmsted, Frederick L., Jr., 24, 55
Olmsted, Marion, 24
Ontario Department of Agriculture, forestry bureau, 146
Oregon, forests, 13; state forester, 193; lumbering, 194–196
Oregon City (Ore.), paper co., 194
Ottawa (Can.), forestry congress, 144–146
Owenbey, Mrs. Jerusha, 63
Oxford University, forestry instruction, 5

PACIFIC COAST PIPE CO., 197
Pacific Creosoting Co., 197
Pacific Logging Congresses, Biltmore students at, 198
Pacific Northwest, Biltmore school visits, 193–197
Parcus, Carl, German banker, 18
Peed, W. W., at logging congresses, 198
Pegu (India), forestry, 9–11

Pennsylvania, forestry, 13, 22
Pennsylvania Forestry Assn., secretary, 22
Pennsylvania R.R., catalpa plantations, 189
Peters, J. Girvin, Biltmore lectures, 152
Petersham (Mass.), forestry school, 168
Pettis, Clifton R., heads forest nursery, 186
Phayre, Arthur, 11
Pheasant hatcheries, Biltmore, 60, 137
Philippine Islands, forests regulated, 141
Phillips, Karl, silviculture system, 183
Pierson, Albert H., Biltmore student, 80
Pigeon River, tract sold, 52; tree species, 52; splash dams, 187; logging: 187, contracts, 59
Pillsbury, John S., gov. Minnesota, 72, 73
Pinchot, Gifford, 23, 33, 54, 65, 72, 80, 154, 184; influenced by Brandis, 8; private forests working plan, 12, 14; relations with Sargent, 15, 48, with Schenck, 19–21, 51, 78, 88, 117–121, 132, 133, 165, with Fernow, 109, with Roosevelt, 111, 112; Biltmore forester: 18, 21, 25–27, 44, 45, 67, 72, 76, resigns, 31; characteristics and interests, 19, 20, 31, 32; visits Biltmore, 28–31, 49, 51; author, 31n., 43; sec. forest commission, 31n., 47, 48; heads U.S. Div. of Forestry, 31n., 78, 86, 87, 112; Pisgah Forest plans, 34, 35, 40, results, 50; advises Pisgah purchase, 41; Webb estate plans, 43; plan for forest commission, 43; views on forest reserves controversy, 49; encourages establishment of Biltmore school, 77; forestry in Nehasane, 87, 88, 93; establishes Yale school, 92; views on forestry, 122, 132, 133, 139, attacks lumbermen, 132; founds foresters' society, 138; results of forestry propaganda, 141
Pinchot, James W., 19, 20
Pinchot, Mrs. James W., 19
Pinkbeds, 63, 103; interior holdings, 30, 64, 65; mountaineers 63–66; roads, 65; church, 66, 80, 134; log yard, 67; deer hunt, 100; Schenck's

headquarters, 127, 156, 171; store, 134; logging operations, 171. See also Pisgah Forest
Pisgah Forest, 13, 113, 119; working plan, 12, 81–85, 104; described, 28, 30; roads and trails, 29, 34, 37, 59, 67, 84, 85, 121, 129, 133, 137, 171, 193; tree species: 29, 36, 40, 42, 101, 135, 145, 149, inventory, 81–83; mountaineers, 30, 58, 63–67, 172; maps, 33, 34; logging: 37, 40–42, 49, 50, 53, 163, 167, proposed, 28, 29, lawsuits resulting, 28, 29; afforestation, 49, 50, 97, 163; acreage, 51; rangers, 51, 121; interior holdings, 52, 84, 85, 129, 130, 137; hunting lodge, 58, 59, 99, 100, 127; search for gold, 70, for radium, 71; mica mine, 71; forestry criticized, 73; resources: 82, 83, tannic wood, 104, 105; investment and revenue estimated, 84, 85; fires, 97, 171; railroad facilities, 103, 105, 129, 134; lumberyard, 129; population, 129; sawmills, 129; school working field, 133, 156, 174; bridges, 134; stores, 134; mail service, 134, 175; pheasant hatchery, 137; sportsmen's leases: 146, 147, 172, estimated, 84; pulpwood, 148; proposed sale, 155, 156; Schenck's property in, 174, 187, 188; school, 175; Schenck's accomplishments in, 175, 188; revisited by Biltmore school, 188; sold, 207; under U.S. Forest Service, 208. See also Big Creek; Forest department; Pinkbeds
Pisgah Ridge, 128; pastures, 28; Pinchot's hut, 30
Polson Logging Co., 196
Porter, Herbert K., Biltmore student, 80
Portland (Ore.), Biltmore school visits, 193, 194; logging co., 198
Potlatch Lumber Co., 198
Potter, Albert F., 186
Price, Overton, influenced by Brandis, 8; assists Brandis, 14; forestry apprentice, 43, student, 80, 117; Pinchot's assistant, 80, 117; member foresters' society, 139
Prussia (Germany), Feldjägers, 119
Public Lands Commission, report on federal land laws, 140

QUEBEC (Can.), forestry meeting, 139

RANGERS. *See* Forest rangers
Rankin, Cyrus T., ranger, 60
Red River Lumber Co., 73
Reed, Franklin W., assists Schenck, 129
Rees, Hans, tanner, 153
Reeves, Ulysses, ranger, 64
Ribbentrop, Baron Bernhardt von, 154; heads forestry in India, 1, 1n., 41; visits Biltmore, 40
Riordan, Dan M., visits Biltmore, 39
Roads, Biltmore: 46, 75, 90, Pisgah Forest, 29, 34, 37, 59, 65, 67, 84, 85, 121, 129, 137, 157, 171, 193; importance to forestry, 47, 67
Robertson, Reuben B., manager fibre co., 148, 150, 187
Roentgen, Wilhelm K., physicist, 3
Roosevelt, Franklin D., political versatility, 102
Roosevelt, Theodore, praises Brandis, 8; calls forestry congress, 101; succeeds to presidency, 111; relations with Pinchot, 111, 112; visits Biltmore, 112, 117; speaker, 112, 142; attitude toward lumbermen, 118, 132, 142; appoints Public Lands Commission, 140; results of forestry propaganda, 141
Roper (John L.) Lumber Co., 186, 187; host to Biltmore school, 202
Ross, Malcolm, farm dept. asst., 92
Ross, Norman, heads Canadian nurseries, 146
Roth, Filibert, timber physics expert, 22
Rothrock, Joseph T., 47; sec. forestry assn., 13, 13n., 22; originates Pennsylvania forestry movement, 22; editor, 22
Royal Indian Engineering College, graduates tour European forests, 4–7, 10; founder, 16

ST. CLOUD (France), forests, 184
St. Helens (Ore.), creosoting plant, 194
St. Joe National Forest, Biltmore school visits, 197
St. Maries (Idaho), lumber co., 167
St. Paul and Tacoma Lumber Co., host to Biltmore school, 196
Santa Clara Lumber Co., 186
Saranac Lake (N.Y.), fish hatchery, 186
Sargent, Charles S., author, 14, 15, 22;

relations with Pinchot, 15, 48; arboretum director, 22; editor, 22, 48; visits Biltmore, 36, 37; plan for forest commission, 43; heads commission, 47, 48; views on forest reserves controversy, 48; relations with Fernow, 109
SARGENT, CHARLES S., *Silva of North America*, 14, 14n., 15, 22; *Report on Forests of North America*, 14, 14n., 22
Sargent, John S., portrait painter, 98
Saunders, William L., lumber co. manager, 189
Sawmills, Asheville, 24, 38, 41, 44, 49; Hickory, 37; Jackson County, 38; Pisgah Forest, 63, 129; Transylvania County, 67; Sunburst, 187; Coos Bay (Ore.), 194
Saxony (Germany), Brandis' notes on forests, 14
Schenck, Carl A., background and education, 1–7, 16; Biltmore forester, 1, 17, contract, 18, 33, 57, 94, 131, 173; assists Brandis, 5–7, 10, 14, 183; German military service, 16, 57, 77, 94–96, 131; mother, 18, 139; first impressions of U.S., 18–20; relations with Pinchot, 19–21, 51, 78, 88, 117–121, 132, 133, 165, with Fernow, 21, 22, 109, 110, with Vanderbilt, 32, 33, 80, 85, 128, 133, 155, 172, 188; author, 22, 23, 117, 175–177, 211, 212; Bltmore office, 23, 126, residence, 33, 36, 37, 93; views on forestry, 43, 44, 50, 98, 121–123, 125, 132, 133, 139, 141, 166, 175, 176, 203–207, on forest reserves controversy, 49, on education, 169, 170; marriage, 57; trips to Europe, 57, 94–96, 132, 154, 160, with students, *see* Biltmore Forest School; speaker, 62, 81, 87, 87n.; 123, 124, 132, 135, 142–144, 166; Pisgah headquarters, 63, 149; resigns as lumber co. pres., 69; visits Minnesota, 72–75, 88–90; report on forestry for Minnesota, 74, 75; establishes Biltmore school, 77; with German Forest Service, 77, 94, 130; consulting forester, 78, 80, 81, 113, 115, 167, *see also* Schenck (C. A.) and Co.; agent U.S. Div. of Forestry, 78, 86, 87; leads church services, 80, 134; offered professorship at Giessen, 94; citizenship, 113, 117; director Hard-

wood Manufacturers' Assn., 135; assists Schlich, 138; sister, 139; member foresters' society, 139; at Canadian forestry meetings, 139; forest congress address quoted, 142–144; visits Michigan, 158–160; encounter with Beadle, 171, 172; dismissed by Vanderbilt, 173; Pisgah property, 174, 187, deeded to Vanderbilt, 188; Biltmore accomplishments summarized, 175–177; asst. professor at Darmstadt, 177; sues Vanderbilt, 188; leaves U.S., 201; farewell message to Biltmore students, 201–207. *See also* Biltmore Estate; Biltmore Forest; Biltmore Forest School; Forest department; Pisgah Forest

Schenck, Mrs. Carl A., 17, 97, 98, 100, 167; marriage, 57; as forester's wife, 58; relations with mountaineers, 58, 66, with Mrs. Vanderbilt, 79, 80; European trips, 96, 139; entertains Biltmore guests, 156, 168; hurricane experience, 191

Schenck (C. A.) and Co., consulting foresters, 114, 114n., 167; activities, 114–116, 150

Schenck, Heinrich, 180

Schlich, Sir William, 154; in Indian Forest Service: 5, 12, inspector general, 4, 5, 40, develops forestry school, 10; views on forestry, 11; assists Brandis, 14; founds engineering college, 16; European forestry tours, 16, 96, 138

SCHLICH, WILLIAM, *Manual of Forestry*, 51

Schönmünzack (Germany), state forests, 182, 183

Schrenk, Hermann von, visits Biltmore, 138; Biltmore school lectures, 190

Schwarzwald Forest (Germany), 14

Scott, Charles F., visits Biltmore, 157

Seattle, (Wash.), 194; Biltmore school visits, 197

Seattle Cedar Manufacturing Co., 197

Seton, Ernest Thompson, Biltmore school lectures, 190, 191

Sewanee (Tenn.), university, 81

Sherman, Franklin, Biltmore lectures, 152

Siebert, John O., advises Schenck, 17

Silviculture, combined with agriculture, 10, 180, 183. *See also* Forestry

Simpson Logging Co., 196

Sioussat, St. George L., Biltmore lectures, 92

Smith, Charles A., pres. lumber co., 194

Smith (C. A.) Lumber Co., hosts to Biltmore school, 194, 202; logging, 195; fibre plant, 196

Smoot, Reed, visits Biltmore, 158

Society for the Protection of the Adirondacks, 108

Society of American Foresters, 138, 139

Sorrels, Lafayette, ranger, 30, 63

Southern Lumberman, editor, 124

Southern Railroad, 38; acquires feeder lines, 103

Southern states, pitch pine forests, 13; pine forests bought, 53; hardwoods business, 68; pines, 87; timber canvass advocated, 117

Spessart Forest (Germany), 6, 14

Splash dams, 39; Germany, 29; Big Creek, 29, 35, 37, 40–42, north fork, 42; Hickory, 37; Pigeon River, 187

Steigerwald Forest (Germany), 14

Sternburg, Speck von, ambassador to U.S., 101

Steuben, Baron Richard von, forester, 21

Stuart, Melvin N., miller, 64

Sudworth, George B., heads Bureau of Forestry division, 112

Sugarloaf Mountain, corundum mine, 113

Sunburst, village built, 150; Biltmore school headquarters, 187

Swannanoa River, 126; nurseries, 24; plowlands, 26

Sweden, forestry, 72

Switzerland, Brandis' tours, 5; Biltmore tours, 183, 184

Swope, Dr. Rodney R., Biltmore pastor, 168

Syracuse (N.Y.), forestry school, 177

TACOMA (Wash.), Biltmore school visits, 196

Taft, William H., elected president, 161, 166; views on lumber tariff, 166

Tanbark and tannic wood, Pisgah Forest, 103–106

Tannic-acid plant, Brevard, 103, 104, 149

Tanning industry, North Carolina, 121, 187

Tariff, lumber, 165, 166; Schenck's views, 176

Taunus Mountains (Germany), sanatorium, 3

Tennessee, forest fire laws, 115; efforts for forestry assn., 123

Tennessee Copper Co., lawsuit, 114

Thaer, Albrecht, agriculture teacher, 3

Thomson, Peter G., pulp factory project, 148–150

Three Day Camp, 127, 128

Three State Lumber Co., 171, 172

Thun (Switzerland), forests, 183

Timberman, editor, 193, 198

Toxaway, Lake, railroad to, 134

Transylvania County, 65, 67, 103

Tuckaseegee River, forest, 113

Tupper Lake (N.Y.), Cornell school laboratory, 93; Biltmore school headquarters, 178, 186

Twin Falls Logging Co., 198

UNITED STATES, ambassadors to, 99, 100, 101, 102

U.S. Biological Survey, 152

U.S. Bureau of Entomology, 152

U.S. Bureau of Forestry, 118, 119, 141; divisions, 112; established, 112, 199

U.S. Congress, members visit Biltmore, 157, 158

U.S. Department of Agriculture, forest reserves transferred to, 140. *See also* U.S. Bureau of Forestry; U.S. Division of Forestry

U.S. Department of the Interior, forest reserves transferred from, 140

U.S. Division of Forestry, 78, 80; budget, 21, 79; created, 199

U.S. Forest Service, 152, 165; timber canvass of South, 117; attitude toward lumber tariff, 176; criticized, 207; manages Pisgah Forest, 208

U.S. Geological Survey, Pisgah Quadrangle, 34

U.S. Reclamation Service, chief, 92

U.S. Shipbuilding Co., 128, 131

U.S. Supreme Court, Georgia-Tennessee suit, 114

University of Giessen, 16; forestry school, 3, 4; visited by Brandis' tours, 4, 5; offers Schenck professorship, 94

University of Minnesota, 72; tree nursery, 73

University of North Carolina, 91, 92

University of Ohio, Schenck's lecture, 87

University of the South, 92, 124; Schenck's plan for woodlands, 80, 81

University of Tennessee, 123

University of Toronto, forestry dept., 146; forestry dean, 151, 152, 160

University of Tübingen, 4; forestry school, 2; students' club, 3

University of Washington, forestry school, 197

VANDERBILT, CORNELIA, 98

Vanderbilt, Cornelius, 32

Vanderbilt, George W., 23, 88, 93, 100, 190, 202, 208; engages Schenck, 1, 2, 17; contract with Schenck, 18, 33, 57, 94, 131, 173; buys Pisgah Forest, 28; characteristics and interests, 32; relations with Schenck, 32, 33, 80, 85, 128, 133, 172, 188; instructions *re* Big Creek logging, 34; orders re-establishment of hardwood forest, 45, 55; approves Schenck's plans, 46, 77, 85, 104; considers Pigeon River purchase, 52; opposes workmen living on estate, 59; visits Pisgah, 67, 80, 127, 128; shares Schenck's professional fees, 78, 113, 167; marriage, 79; portrait, 98; interest in building, 126–128, in landscaping, 170; invests in shipbuilding co., 128; financial losses, 131, 133, 137, 146, 155; absences from Biltmore, 147, 154, 156; Biltmore project compared with Thomson's, 149; orders sale of Pisgah, 155; dismisses Schenck, 173; seizes Schenck's Pisgah property, 187; sued by Schenck, 188, 189; death, 207

Vanderbilt, Mrs. George W., 98, 156; arrives at Biltmore, 79; relations with estate residents, 79; visits Pisgah Forest, 80; sells Pisgah, 207, 208

Vanderbilt, William H., 32

Vanderbilt, Mrs. William H., 32

Vanderbilt, William K., 32

Versailles (France), forests, 184

Vinnedge, R. W., at logging congresses, 198

WAGNER, CHRISTOPH, author, 182

Walker, Thomas B., lumberman, 73

Walker County (Ala.), longleaf pine study, 86
Wang, Wu, ambassador to U.S., 99
Washington (D.C.), Biltmore school visits, 186
Waterbury, John I., pres. trust co., 147
Webb, W. Seward, 32; Adirondacks estate, 43, 87, 88. *See also* Nehasane Park
Webb, Mrs. W. Seward, 32, 100
Weeks Law, importance to forestry, 199
Western Crossarms Co., 197
Western Forestry and Conservation Assn., 198
Weston, George F., heads farm dept., 24, 137
Weyerhaeuser, Frederick, forest holdings, 53; on forestry board, 72, 75
White, T. Stewart, host to Biltmore school, 189
Whitney, Charles L., logging operations, 25, 29, 34, 37, 41

Wilhelm II, German Kaiser, 177
Willamette Iron and Steel Works, 194
Willamette River Valley, mills, 194
Wimmenauer, Karl, forestry teacher, 3
"Woodcot," Schenck's Biltmore residence, 33, 36, 37, 93
Woodcutting plant, Biltmore, 60
Württemberg (Germany), forests and forestry, 13, 14, 182

YALE SCHOOL OF FORESTRY, 97, 132; established, 92; lacks working field, 93; students visit Biltmore, 137; library, 176
Yale University, plans forestry course, 81; forestry chair, 92. *See also* Yale School of Forestry
Young, J. D., at logging congresses, 198
Ysenburg (Germany), forests, 181

ZURICH (Switzerland), Sihlwald town forest, 184